AIoT and Smart Sensing

AIoT and Smart Sensing: A Comprehensive Guide to the Next Generation of Smart Devices offers an in-depth exploration of the intersection of Artificial Intelligence of Things (AIoT) and smart sensing technologies. As the convergence of AI and IoT reshapes industries, this book serves as an essential guide for understanding the technological foundations, security protocols, and wide-ranging applications that make AIoT a transformative force. By examining both foundational and applied aspects, this book aims to provide readers with a holistic view of how AIoT is driving innovation in agriculture, healthcare, smart cities, and beyond.

What sets this book apart is its dual focus on technological frameworks and real-world applications. The first part addresses key security issues, technological innovations, and practical implementations. The second part demonstrates AIoT's impact on diverse sectors, including agriculture, healthcare, and cultural fields. By linking theory with practice, this book not only introduces cutting-edge concepts but also showcases their potential for revolutionizing industries.

Key features include:

- Comprehensive coverage of AIoT security protocols, including RFID systems, blockchain in healthcare, and multi-cloud environments in smart cities.
- Detailed case studies on precision farming, AI-driven crop management, and sustainable agriculture.
- Exploration of AI innovations in medical diagnostics, chronic healthcare management, and personalized patient care.
- Unique cultural applications, such as AI-based recognition of Carnatic ragas, highlighting AIoT's versatility.
- Future trends in AIoT for healthcare, including advanced monitoring and diagnostic systems.

This book is designed for a wide audience, including researchers, professionals, and students in fields such as AI, IoT, healthcare, agriculture, and smart city development. It is an invaluable resource for anyone seeking to understand the future of smart sensing and AIoT-driven technologies.

AIoT and Smart Sensing
A Comprehensive Guide to the Next Generation of Smart Devices

Edited by
Vaishali R. Kulkarni, Thompson Stephan,
Punitha S, Fadi Al-Turjman,
and Thinagaran Perumal

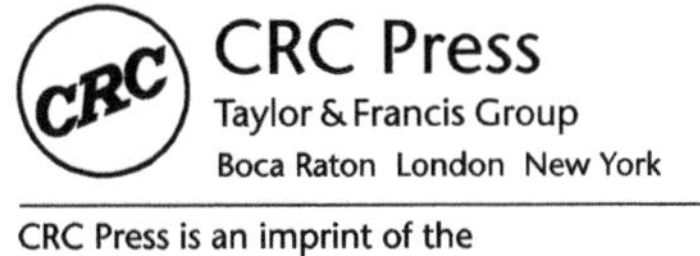

CRC Press
Taylor & Francis Group
Boca Raton London New York

CRC Press is an imprint of the
Taylor & Francis Group, an **informa** business

Contents

Part 2
Applications and future directions

About the editors

Dr. Vaishali R. Kulkarni earned her B.E. degree in Computer Science and Engineering from Karnataka University, Dharwad, India, and her M. Tech. and Ph.D. degrees in Computer Science and Engineering from Visvesvaraya Technological University, India. Her research interests include wireless sensor networks, sensor node localization, swarm intelligence, fuzzy logic, and neural networks. She is an active member of IEEE, a life member of the Indian Society for Technical Education (ISTE), and a member of the Computer Society of India (CSI). Dr. Kulkarni has served as an invited reviewer, session chair, and author at international conferences and journals. Her work includes numerous book chapters, refereed journal articles, and conference papers.

Dr. Punitha S is Associate Professor in the Department of Computer Science and Engineering at Graphic Era Deemed to be University, Dehradun, Uttarakhand, India. She holds a Ph.D. in Computer Science Engineering from Pondicherry Central University, Puducherry, India, with a focus on advanced diagnostic techniques for breast cancer using metaheuristic optimization. Her research interests include evolutionary computation, machine learning, and artificial intelligence. Dr. Punitha has published extensively in high-impact journals and conferences, particularly in the areas of medical diagnostics, AI, and optimization techniques. She has also co-authored several book chapters and has been actively involved in research projects.

Dr. Thompson Stephan earned his PhD from Pondicherry University, India, in 2018, has nearly seven years of academic experience, complemented by full-time research and industry expertise. He serves as an Assistant Professor at Thumbay College of Management and AI in Healthcare, Gulf Medical University, Ajman, United Arab Emirates. Recognized among Stanford/Elsevier's Top 2% Scientists globally in both 2023 and 2024, Thompson has received prestigious accolades, including the Best Researcher Award in 2020 and the Protsahan Research Award in 2023, both from the IEEE Bangalore Section, India. His primary research focus is in artificial intelligence, with specialized expertise in advancing machine learning, data mining, and metaheuristic optimization. With more than 80 Scopus-indexed publications, including 48 in SCI-indexed journals, Thompson's work has garnered significant recognition. He actively contributes as a book editor and reviewer for esteemed international journals, with publications on leading platforms such as IEEE, Elsevier, Taylor & Francis, and Springer.

Prof. Dr. Fadi Al-Turjman received his Ph.D. in Computer Science from Queen's University, Kingston, Ontario, Canada, in 2011. He is the advisor to the Chairman of the Board of Trustees on AI and Informatics and Associate Dean for Research at Near East University (NEU), Nicosia. He also heads the Artificial Intelligence, Information Systems, and Software Engineering Departments and directs the AI and Robotics Institute and the International Research Center for AI and IoT at NEU. Prof. Al-Turjman has received the 2022 Lifetime Golden Award of Dr. Suat Gunsel from NEU, among other prestigious accolades and best research awards. A leading expert in smart IoT systems, wireless, and mobile networks, he has published over 600 SCI/E papers and authored/edited more than 70 books with top publishers, including IEEE, Taylor & Francis, Elsevier, IET, and Springer. He frequently delivers keynotes and plenary talks at international conferences and serves as a guest/associate editor for leading journals such as *IEEE Communications Surveys and Tutorials* and Elsevier's *Sustainable Cities and Society.*

Prof. Dr. Thinagaran Perumal, a recipient of the 2014 Early Career Award from the IEEE Consumer Electronics Society for his pioneering work in consumer technologies, completed his Ph.D. at Universiti Putra Malaysia in the area of smart technology and robotics. He is currently Head of the Department of Computer Science at Universiti Putra Malaysia, Selangor Darul Ehsan. His research focuses on the interoperability of smart homes and the Internet of Things (IoT), wearable computing, and cyber-physical systems. Prof. Perumal chairs the National Committee on Standardization for IoT and leads research in cognitive IoT frameworks for smart homes and wearable devices for rehabilitation. He is an active member of the IEEE Consumer Electronics Society and serves on the Future Directions Committee on IoT. He has published extensively in IEEE conferences and journals and holds leadership roles in several prominent IEEE conferences, including serving as General Chair of the 2017 IEEE International Symposium on Consumer Electronics. Prof. Perumal is currently Vice Chair of the IEEE CTSoc Technical Committee on IoT and Digital Twin, overseeing the society's technical activities.

Contributors

Fadi Al-Turjman
Artificial Intelligence, Software,
and Information Systems
Engineering Departments, AI
and Robotics Institute
Near East University
Nicosia, Turkey

Balaji Arulraj
Division of Data Science and Cyber
Security
Karunya Institute of Technology
and Sciences
Coimbatore, India

Aneesha M. Bail
M. S. Ramaiah University of
Applied Sciences
Bangalore, India

Almas Begum
Department of Computer Science
and Engineering, Saveetha
School of Engineering
Saveetha Institute of Medical and
Technical Sciences
Chennai, India

Carmel Mary Belinda
Department of Computer Science
and Engineering, Saveetha
School of Engineering
Saveetha Institute of Medical and
Technical Sciences
Chennai, India

Rakshita Bhardwaj
Department of Computer Science
and Engineering
M. S. Ramaiah University of
Applied Sciences
Bangalore, India

P. PadmaPriya Darshini
M. S. Ramaiah University of
Applied Sciences
Bangalore, India

S. Alex David
Department of Computer Science
and Engineering
Vel Tech Rangarajan
Dr. Sagunthala R&D Institute of
Science and Technology
Chennai, India

H. Deepika
M. S. Ramaiah University of
    Applied Sciences
Bangalore, India

Joy Winston James
Department of Computer Science
    and Engineering, University of
    Computer Studies
University of Technology
Salmabad, Bahrain

Meenalochini Jeganathan
Department of Geography
M. V. Muthaiah Government Arts
    College for Women
Dindigul, India

P. Rajesh Kanna
Department of Computer Science
    and Engineering
Bannari Amman Institute of
    Technology
Sathyamangalam, India

E. Kannan
Department of Computer Science
    and Engineering
Vel Tech Rangarajan
    Dr. Sagunthala R&D Institute of
    Science and Technology
Chennai, India

Rachana Reddy Koppala
Department of Computer Science
    and Engineering
M. S. Ramaiah University of
    Applied Sciences
Bangalore, India

K. Muthumanickam
Department of Information
    Technology
Kongunadu College of Engineering
    and Technology
Tholurpatti, India

Mridula G. Narang
M. S. Ramaiah University of
    Applied Sciences
Bangalore, India

P. Pandiaraja
Department of Computer Science
    and Engineering
Vel Tech Rangarajan
    Dr. Sagunthala R&D Institute of
    Science and Technology
Chennai, India

Punitha S.
Department of Computer Science
    and Engineering
Graphic Era (Deemed to be
    University)
Dehradun, India

Deepa Rajamanickam
Department of Computer Science
    and Engineering
Vels Institute of Science,
    Technology & Advanced Studies
Chennai, India

**Rajalakshmi N. Nagarnaidu
Rajaperumal**
Department of Computer Science
    and Engineering
Vel Tech Rangarajan
    Dr. Sagunthala R&D Institute of
    Science and Technology
Chennai, India

Gnanajeyaraman Rajaram
Department of Computer Science
    and Engineering, Saveetha
    School of Engineering
Saveetha Institute of Medical and
    Technical Sciences
Chennai, India

Manoranjitham Rajendran
Division of Artificial Intelligence
  and Machine Learning
Karunya Institute of Technology
  and Sciences
Coimbatore, India

Ramiz Salama
Department of Computer
  Engineering, AI and Robotics
  Institute, Research Center for AI
  and IoT
Near East University
Nicosia, Turkey

K. Senthil
Department of Computer Science
  and Engineering, Saveetha
  School of Engineering
Saveetha Institute of Medical and
  Technical Sciences
Chennai, India

**Supriya Mallanayakanakatte
Shankaregowda**
Department of Computer Science
  and Engineering
M. S. Ramaiah University of
  Applied Sciences
Bangalore, India

H. N. Sharul
M. S. Ramaiah University of
  Applied Sciences
Bangalore, India

S. Sivagami
Department of Computer Science
  and Engineering, Saveetha
  School of Engineering
Saveetha Institute of Medical and
  Technical Sciences
Chennai, India

K. Sumathi
Department of Computer Science
  and Engineering
Builders Engineering College
Kangeyam, India

R. S. Archana Vishveswari
Department of Computer Science
  and Engineering
Varuvan Vadivelan Institute of
  Technology
Dharmapuri, India

Technological foundations and innovations

Part 1 lays the groundwork for understanding Artificial Intelligence of Things (AIoT) and smart sensing, focusing on the technological frameworks and security protocols that ensure system integrity. It introduces the reader to key innovations in AIoT, emphasizing the importance of robust security measures across various sectors. The chapters explore the critical role of security in AIoT systems, from advanced anomaly detection to multi-cloud environments in smart cities. Part 1 also addresses how blockchain technology enhances cybersecurity in healthcare, followed by practical applications in agriculture, including precision farming and AI-driven decision support systems. This foundation equips readers with the knowledge needed to appreciate the transformative potential of AIoT.

DOI: 10.1201/9781032618173-1

## Chapter 1

Advanced security

AI-powered anomaly detection for RFID door locks using Arduino and FreeRTOS

Balaji Arulraj, Manoranjitham Rajendran, and Punitha S

1.1 INTRODUCTION

In today's tech-driven world, there is a profound shift in how individuals approach even the most basic aspects of their daily routines. Access control systems are at the forefront of this major revolution in security in particular. More advanced locks that meet the needs of today's society are progressively replacing traditional locks. With an emphasis on the ground-breaking radio frequency identification (RFID) door lock solution enhanced with AI-based anomaly detection, this chapter sets out to explore this fascinating evolution. The days of struggling with traditional keys and trying to remember complicated passwords are long gone. This system transforms access management by utilizing RFID technology. Through a smooth integration of FreeRTOS with Arduino microcontrollers, a system has been created that prioritizes user ease while strengthening security. Imagine being able to pass through doors with just a card or tag swipe, all the while knowing that state-of-the-art technology, including AI algorithms, is protecting the area.

However, the goals go beyond practicality. The guiding principle is that security ought to be adaptable, expandable, and available to all. Whether securing apartment buildings, office buildings, or academic institutions, this RFID door lock solution offers a flexible framework that can be tailored to suit the specific requirements of any given environment. The complexities of this system, from conception to implementation, will be negotiated. The technical challenges encountered and the creative solutions devised to overcome them will be discussed. Additionally, real-world examples and useful applications will be highlighted, demonstrating how the technology can actually improve security and expedite processes.

RFID technology is recognized for its cost-effective and efficient nature, playing a significant role in enhancing security measures and access control. A digital security system integrating RFID technology for door locking provides features such as real-time authentication and user validation. This ensures secure access while maintaining comprehensive logs of user activities, enhancing overall security. Passive RFID technology offers advantages like affordability and compact size, making it suitable for a variety

DOI: 10.1201/9781032618173-2

of applications. These attributes contribute to its versatility and practicality in different settings, ranging from residential to commercial environments. Overall, RFID technology demonstrates a robust solution for improving digital security, showcasing its effectiveness in modern access control systems [1].

A key feature of this solution is the integration of k-nearest neighbours (k-NN) algorithms for real-time anomaly detection. This AI component monitors access patterns and environmental data to detect and alert any unusual behaviour, significantly enhancing the security of the system. The k-NN algorithm, known for its simplicity and effectiveness in pattern recognition, continuously learns from access data, making it proficient in identifying anomalies that may indicate security breaches or unauthorized access attempts. The system's ability to adapt and improve its detection accuracy over time ensures a proactive security stance.

In addition, this chapter will explore the economic implications of adopting advanced access control systems. The cost-effectiveness of implementing RFID door lock solutions compared to traditional locking mechanisms, considering factors such as installation, maintenance, and long-term operational costs, will be analysed. By providing a comprehensive cost–benefit analysis, decision-makers can assess the value proposition of upgrading their security infrastructure. Moreover, the potential challenges and opportunities associated with the widespread adoption of RFID door lock solutions will be delved into. From regulatory compliance to interoperability with existing systems, key considerations that organizations may encounter on their journey towards modernizing access control will be addressed. By proactively identifying and addressing these challenges, a smoother transition to the next generation of security solutions can be ensured.

Through collaboration, innovation, and a commitment to excellence, it is believed that the RFID door lock solution with AI anomaly detection has the potential to redefine access control in the digital age. By sharing insights, experiences, and lessons learned, it is hoped to inspire others to embrace innovation and a future where security is not just a necessity but a seamless and integral part of everyday life. Furthermore, the system's AI capabilities can be expanded to include predictive analytics, providing users with insights into potential security risks before they manifest.

Finally, the accessibility features of RFID door lock solutions will be explored, aiming to empower individuals with disabilities and foster inclusivity and equal access to spaces in communities. Enhancing user interfaces and integrating voice commands or other assistive technologies can make these systems more accessible. Future iterations may also incorporate biometric authentication methods such as fingerprint or facial recognition, adding an additional layer of security and personalization. The potential for integrating with IoT devices to create a comprehensive smart home or smart building ecosystem will also be discussed, highlighting the possibilities for future advancements in this technology.

By pushing the boundaries of what is possible with RFID technology and AI, this research sets the stage for more secure, efficient, and user-friendly access control solutions that meet the diverse needs of modern users.

1.2 LITERATURE REVIEW

Komol et al. (2018) present a comprehensive study on RFID to create a robust dual security system for access control. The authors highlight the increasing importance of automated access control systems in mitigating security threats faced by organizations, particularly in safeguarding sensitive items and areas. Traditional methods such as passwords and ID cards are prone to hacking and loss, necessitating more reliable alternatives. RFID technology emerges as a promising solution due to its wireless data transmission capabilities and versatility across various sectors. However, security concerns arise due to vulnerabilities in the wireless channel between RFID tags and readers. To address these challenges, the authors propose the integration of fingerprint biometrics, a well-established and widely accepted form of biometric authentication [2]. Security concerns have been paramount in various domains, prompting the development of advanced control systems to prevent unauthorized access. Traditional lock-and-key mechanisms, while widely used, have limitations in terms of practicality and susceptibility to security breaches. Consequently, researchers have explored alternative approaches integrating electronic identification systems. One such technology gaining prominence is RFID, which enables wireless data transmission for automatic identification. Despite initial challenges related to standardization, RFID technology has emerged as a reliable and cost-effective solution for access control systems. Moreover, the use of proximity sensors adds an additional layer of protection by detecting the status of the door, whether open or closed. This real-time sensing capability enables prompt notifications to users via email, providing them with timely information about potential control with door status sensing; the proposed system offers comprehensive security coverage for residential and office environments [3]. The integration of RFID technology into door locking systems, particularly within the context of IoT-enabled security solutions provides real-time monitoring, allowing for more efficient and secure management of entry points in both residential and commercial settings. It discusses the significance of RFID-based access control systems in various environments, such as hostels, where security monitoring is essential. Additionally, they highlight the widespread adoption of RFID technology across different industries for wireless data transmission, tracking, and access control purposes. It mentions RFID's applications in personal tracking, supply chain management, library book management, tollgate systems, public transport, industrial automation, asset tracking, and more. Overall, they underscore the versatility, efficiency, and security benefits of RFID technology in modern security systems, providing a foundation for the development and implementation of RFID-based door locking systems [4]. Studies on home automation systems with features like speech recognition, face recognition, and internet access highlight the potential for enhanced security, convenience, and energy efficiency, enabling users to control and monitor their home environments remotely and with ease. Some systems have utilized RFID technology for door locking using the ZigBee module, and another system was introduced by Md. Nasimuzzaman Chowdhury et al., which incorporated remote access via the internet. The primary focus is on creating a cost-efficient and readily deployable system for monitoring attendance and ensuring secure access control, with the added capability of customization to accommodate specific needs. It aims to overcome the limitations of existing systems, such as high costs and complexity, by proposing a simple yet effective solution that leverages RFID technology and real-time control mechanisms. Overall, they provide valuable insights into the existing landscape of door locking systems and set the context for the proposed system, which aims to address the shortcomings of previous approaches while offering additional features like real-time monitoring and remote access [5].

Baikerikar et al. [6] introduced a smart door locking system by using one-time password to create a secure and ease of use for the customers [6]. The demerits of this system are high power consumption, power failure, and high cost. Venkatraman et al. [7] proposed LoRa technology for monitoring home security and checking the status of the door. The limitation of this work is it has high bandwidth and requires higher power [7]. AI enhances smart home door locking systems using IoT and machine learning models [8].

1.3 PROPOSED SOLUTION

"RFID door lock using Arduino and FreeRTOS" involves the development of a secure door locking system leveraging RFID technology, Arduino microcontroller, and FreeRTOS for task management. The system architecture comprises an Arduino board interfaced with an RFID reader, a door lock mechanism, and any necessary peripherals. The Arduino sketch will handle RFID tag authentication, door lock control, and RFID reader interfacing, while FreeRTOS will manage concurrent tasks such as RFID scanning, door locking/unlocking, and user interface management. RFID tag authentication will be implemented to ensure secure access, with enrolled tags being authenticated by the system. Error handling mechanisms will be in place to address issues such as unrecognized tags or communication failures with the door lock mechanism. Testing will encompass scenarios to validate RFID tag detection, door locking/unlocking functionality, task scheduling, and error handling. One key aspect will be the implementation of a comprehensive logging and audit trail mechanism to meticulously record all access attempts, providing valuable insights into access patterns and aiding in the detection of potential security breaches.

1.3.1 Motivation of the proposed work

- **Improved Security:** The RFID door lock system enhances security by requiring authorized RFID tags for access, reducing the risk of unauthorized entry. The integration of AI-based anomaly detection further strengthens security by monitoring access patterns and detecting unusual behaviour in real time. This dual-layered approach ensures that any irregularities are promptly identified and addressed, significantly reducing the risk of security breaches.
- **Convenience:** Users can easily gain access without physical keys, enhancing convenience while maintaining security. The AI system ensures that any irregularities are promptly identified and addressed, adding an extra layer of security without compromising user convenience. This eliminates the need to carry bulky keychains and reduces the risk of losing keys.
- **Fast Response:** With FreeRTOS, the system responds quickly to RFID card detection, ensuring prompt access authorization. The AI component operates concurrently to monitor and analyse access events, providing real-time anomaly detection without affecting the system's response time. This ensures that users experience minimal delays when accessing secure areas.
- **Seamless Integration:** The RFID door lock system seamlessly integrates into existing infrastructure, making it easy to deploy and manage within various environments without requiring extensive modifications or disruptions. The AI-based anomaly detection is designed to work with the existing system architecture, adding advanced security features with minimal integration effort. This adaptability makes the system suitable for a wide range of applications.
- **Scalability:** The system is designed to be scalable, allowing for easy expansion as the number of users or access points increases. This is particularly beneficial for organizations that anticipate growth or need to manage multiple access points across different locations.
- **Data-Driven Insights:** The AI component not only enhances security but also provides valuable data-driven insights. Organizations can analyse access patterns to optimize security protocols, improve resource allocation, and enhance operational efficiency.

1.3.2 Application of the proposed work

- **Home Security:** The RFID door lock system can be used in homes to enhance security. It allows homeowners to unlock doors using RFID tags, providing convenient access while ensuring protection against unauthorized entry. The AI-based anomaly detection further improves security by identifying unusual access patterns and alerting homeowners to potential threats. Integration with smart home systems can enable additional features such as remote access and automated locking.
- **Commercial Access Control:** In offices, warehouses, and other commercial settings, the system serves as an effective access control solution.

Employees and authorized personnel can use RFID tags for secure entry, improving workplace security. The AI anomaly detection monitors access behaviour to detect and respond to suspicious activities promptly. This can help prevent theft, unauthorized access, and other security breaches.

- **Education:** Schools and universities can benefit from the system to restrict access to classrooms, labs, and administrative areas. This enhances student and staff safety on campus. The AI component helps to identify and prevent unauthorized access attempts, ensuring a secure educational environment. Additionally, the system can be used to track attendance and monitor the usage of facilities.
- **Hospitality:** Hotels and rental properties can provide secure and convenient access to guests using RFID key cards. This enhances the guest experience and streamlines operations. AI-based anomaly detection adds an extra layer of security by monitoring access patterns and identifying any irregularities. Integration with hotel management systems can further enhance guest services, such as automated check-in and room access.
- **Healthcare:** Hospitals and clinics can control access to patient rooms, labs, and medical storage areas, ensuring patient privacy and equipment security. The AI system detects unusual access patterns, helping to prevent unauthorized entry and protect sensitive areas. Additionally, the system can be used to monitor the movement of staff and patients, ensuring compliance with safety protocols.
- **Government and Military:** Government agencies and military installations can use the system to secure classified areas and restrict access to authorized personnel only. The AI anomaly detection provides an additional security measure by continuously monitoring access events and identifying potential security breaches. The system can also be used to enforce compliance with security policies and track the movement of personnel within secure facilities.
- **Retail:** In retail environments, the RFID door lock system can be used to secure storage areas, cash rooms, and other sensitive locations. AI-based anomaly detection can help identify suspicious activities, such as unauthorized access attempts or unusual patterns of entry, helping to prevent theft and improve overall security.
- **Public Facilities:** Airports, museums, and other public facilities can benefit from the enhanced security provided by RFID door lock systems with AI anomaly detection. These systems can help manage access to restricted areas, ensuring the safety of both staff and visitors. Real-time monitoring and alerts can help security personnel respond quickly to potential threats.

By incorporating these advanced features and capabilities, the RFID door lock system with AI-based anomaly detection offers a comprehensive solution for a wide range of applications, providing enhanced security, convenience, and adaptability to meet the needs of modern users.

1.3.3 Components description

1.3.3.1 Arduino Uno

The Arduino Uno microcontroller board serves as a core component in our RFID door lock system, providing the necessary processing power and interfacing capabilities. As an open-source platform, Arduino Uno leverages the Arduino IDE for firmware development, making it accessible and user-friendly for implementing RFID authentication and door control algorithms. Its affordability, flexibility, and extensive community support make Arduino Uno an ideal choice for our project.

1.3.3.2 RFID reader

In this RFID door lock system, RFID readers play a crucial role in detecting and reading RFID tags or cards presented by users for authentication. These readers operate based on radio frequency signals, communicating with RFID tags to retrieve unique identification data. Use RFID readers integrated with Arduino Uno microcontrollers to facilitate seamless communication and data transfer.

The read range, accuracy, and performance of RFID readers are evaluated to ensure reliable user authentication at door entry points.

1.3.3.3 RFID tag

RFID tags/cards are used for user identification and authentication in this RFID door lock system. These tags come in various types, including passive, active, and semi-passive, each with unique characteristics such as read range and memory capacity. It encodes unique identification data onto RFID tags during manufacturing or programming and registers them with RFID readers to associate them with authorized users in the system database. Security measures, such as encryption and authentication protocols, are implemented to protect RFID tags against cloning and unauthorized access. Practical deployment considerations, including tag placement and durability, are addressed to ensure the effective operation of the RFID door lock system.

1.3.3.4 Servo motor

Servo motors are utilized in this RFID door lock system to actuate the door locking mechanism based on authentication results. These motors operate on feedback control systems, allowing precise positioning of their output shafts. These selected servo motors are suitable for door locking applications and integrate them with Arduino Uno for control purposes. Mechanical design considerations are taken into account to ensure proper mounting and connection of servo motors to door locks, ensuring reliable operation and security.

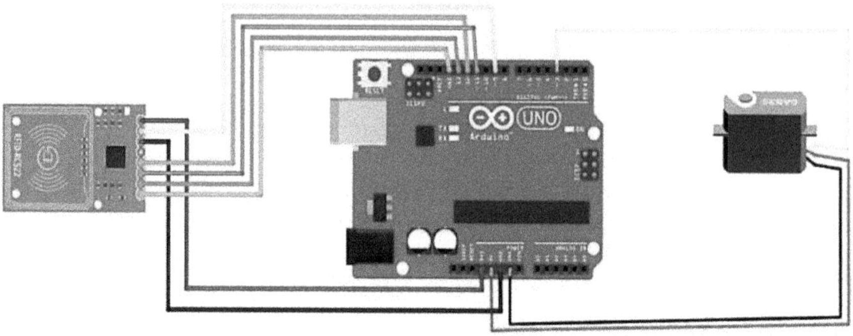

Figure 1.1 Circuit diagram for radio frequency identification (RFID) door lock.

1.3.4 Circuit design

The following are the electronic components which were used to develop this RFID door lock system using Arduino as shown in Figure 1.1.

- Arduino Uno – 1No
- RFID Reader – 1No
- Servo Motor – 1No
- RFID Tag
- Jumpers

Figure 1.2 shows the flow chart representation of the RFID door lock system.

1.3.4.1 Pseudo code

- #include Libraries
- Define Constants
- Define Task Handles
- Setup:
- Initialize Serial Communication
- Initialize SPI
- Initialize MFRC522 RFID Reader
- Attach Servo Motor to SERVO_PIN
- Set initial servo position to close the door
- Create RFIDTask
- Create ServoControlTask
- Loop: (Empty)
- RFIDTask:
- Loop:
- If a new RFID card is present and read successfully:
- Read the UID of the card
- Convert UID to string

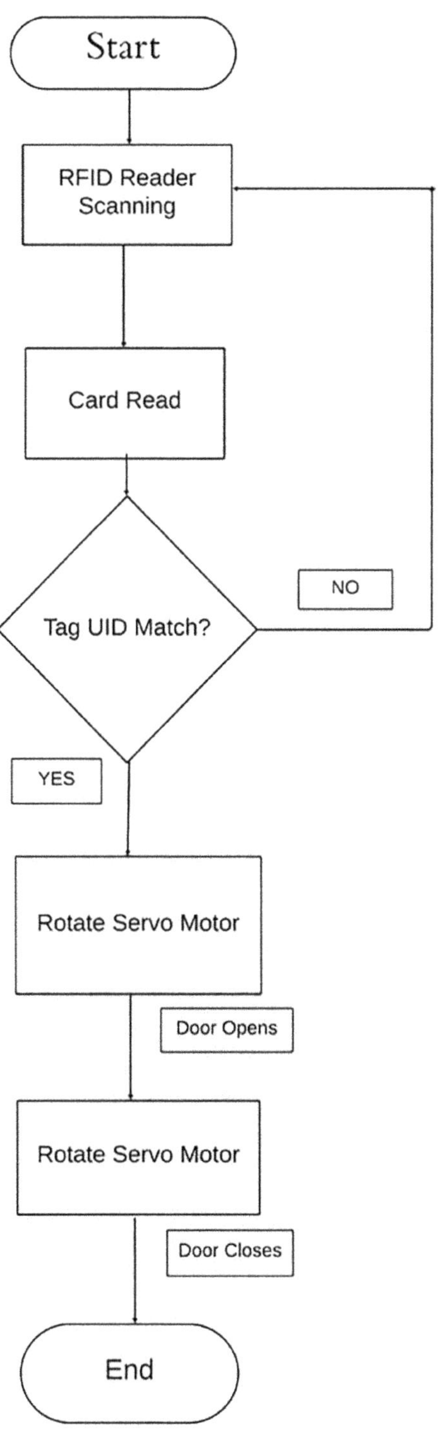

Figure 1.2 Flow chart for radio frequency identification (RFID) door lock system.

- Check if UID matches authorized UID:
- If matched:
  - Print "Authorized access, The Door has Opened"
  - Notify ServoControlTask to open the door
  - Delay for access delay duration
  - Notify ServoControlTask to close the door
- Else:
  - Print "Access denied: Sorry unable to open the door"
  - Delay for denied delay duration
  - Halt the current card session
  - Delay for a short period
  - ServoControlTask:
- Loop:
- Wait for notification from RFIDTask
- Open the door by moving the servo motor to the open position
- Delay for access delay duration
- Close the door by moving the servo motor to the closed position

The front and back sides of the door lock are shown in Figures 1.3 and 1.4. When the authorized UID tag is scanned by the RFID reader, it will display the following shown in Figure 1.5.

When the unauthorized UID tag is scanned by the RFID reader, it will display the following shown in Figure 1.6.

Figure 1.3 Front side of the door.

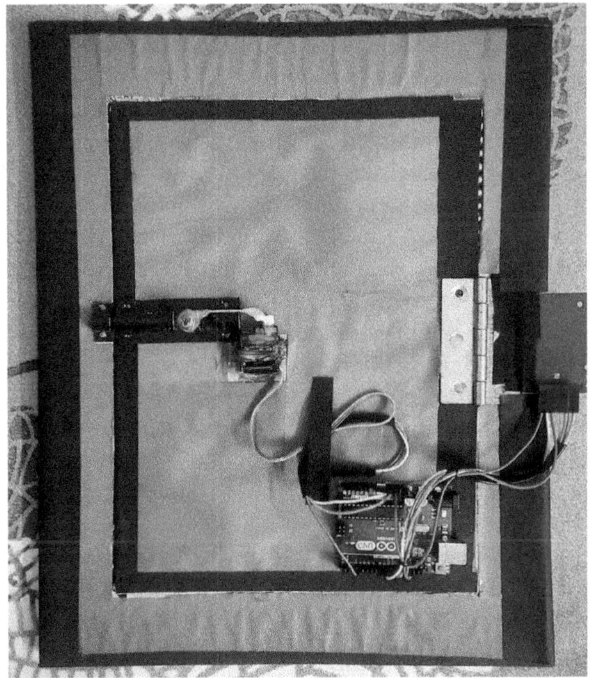

Figure 1.4 Back side of the door.

Figure 1.5 Authorized access.

Figure 1.6 Unauthorized access.

1.3.5 AI anomaly detection

The anomaly detection component of the RFID door lock system plays a pivotal role in enhancing security by continuously monitoring access patterns and identifying irregularities that may indicate potential security threats. The k-NN algorithm is employed for this purpose due to its simplicity and effectiveness in handling real-time data. Unlike traditional rule-based systems that may miss subtle deviations from normal behaviour, the k-NN algorithm analyses the access data by comparing each new access event to a historical dataset of known, legitimate access events [9–11].

The k-NN algorithm operates by calculating the distance between the new data point and all the previous data points in the dataset. It then identifies the "k" closest data points, or neighbours, to the new event. If the majority of these neighbours are classified as normal, the new event is considered normal as well. Conversely, if a significant number of neighbours are classified as anomalous, the system flags the new event as an anomaly. This method is particularly effective in environments where access patterns can vary widely, such as in commercial buildings with fluctuating staff schedules or in multi-residential complexes with diverse occupant routines.

One of the key strengths of the k-NN-based anomaly detection system is its ability to adapt to new patterns over time. As more access data is collected, the model becomes more robust, improving its accuracy in distinguishing between normal and anomalous events. Additionally, the k-NN algorithm's non-parametric nature means it does not make any assumptions about the underlying data distribution, making it versatile and applicable to a wide range of access scenarios.

The real-time nature of this anomaly detection system ensures that any suspicious activity is promptly identified and acted upon. For example, if an RFID tag is used at an unusual time or at an unusual frequency, the system can immediately alert security personnel or trigger automated responses such as locking down the affected area or notifying the user via a mobile application. This proactive approach to security significantly reduces the window of opportunity for unauthorized access, thereby enhancing the overall security posture of the system.

Furthermore, the integration of this anomaly detection model with the RFID door lock system provides a layered security approach. While the RFID technology ensures that only authorized tags can grant access, the k-NN algorithm continuously monitors for any anomalies that might suggest the presence of cloned tags, attempts to bypass the system, or other suspicious activities. This combination of hardware-based and AI-driven security measures creates a more comprehensive and resilient access control solution [12–15].

The k value in a k-NN algorithm is a critical hyperparameter that determines the number of nearest neighbours considered when making a prediction. Essentially, when the k-NN algorithm classifies a new data point, it looks at the "k" closest training data points in the feature space.

The classification of the new data point is then typically determined by a majority vote among these "k" neighbours. For example, if k=3, the algorithm identifies the three nearest neighbours to the new data point, and the new data point is assigned to the class that is most common among these three neighbours [15–17].

Choosing the optimal k value is crucial for the performance of the k-NN classifier. A small k value (e.g., k=1) might make the model sensitive to noise in the training data, leading to overfitting, where the model performs well on the training data but poorly on unseen data. On the contrary, a very large k value might cause the model to be too generalized, potentially underfitting the data, where it fails to capture important patterns.

The k value is optimized using Optuna, a hyperparameter optimization framework. The objective function in the code defines the search space for k to be between 2 and 16. Optuna runs the k-NN algorithm with different k values within this range and evaluates their performance on a validation set. The k value that yields the highest accuracy is selected as the best parameter. This optimal k value is then used to train the final k-NN model. By employing this optimization technique, the goal is to balance bias and variance, achieving the best possible performance on both the training and test datasets.

1.4 EXPERIMENTAL ANALYSIS

1.4.1 Dataset description

The dataset comprises a comprehensive collection of user session data, totalling 34,424 rows and 11 columns, meticulously capturing various aspects of API access behaviours to facilitate anomaly detection. Each entry is uniquely identified by an _id that ensures seamless integration with other datasets if necessary. The dataset includes the inter_api_access_duration(sec), which denotes the time interval in seconds between two consecutive API accesses within a single user session, providing insights into user interaction frequency and potential irregularities. The api_access_uniqueness metric offers a ratio reflecting the diversity of distinct APIs accessed relative to the total number of API calls within the same session, highlighting the variability in user behaviour. Sequence_length(count) indicates the average total number of API calls made in a session, serving as a measure of session activity intensity [18].

Furthermore, the vsession_duration(min) captures the overall length of a user session in minutes within a specified observation window, which is crucial for understanding session longevity and user engagement patterns. The ip_type categorizes the type of IP address used by the user, which can be instrumental in identifying potential security risks associated with certain IP types. Num_sessions records the number of user sessions, each identified by a different session_id, offering a count of session occurrences for each user. Num_users denotes the number of users who generate the same type of API call sequences, providing a gauge for user behaviour commonality. The num_unique_apis counts the distinct APIs accessed in each behaviour group, emphasizing the breadth of API utilization.

The source attribute specifies the origin of the data, ensuring traceability and context for each entry. Finally, the classification column labels each entry as either "normal" or an "outlier," thereby identifying whether the behaviour in the session is considered typical or anomalous. This classification is pivotal for training and evaluating anomaly detection models, such as those using k-NN, to distinguish between regular and suspicious activities effectively. Collectively, these columns offer a rich dataset for analysing API access behaviours, providing valuable parameters for developing robust security measures and anomaly detection systems.

The accuracy of the anomaly detection model is shown in Figure 1.7.

Figure 1.8 shows the performance of a k-NN algorithm. The confusion matrix shows the number of correct and incorrect predictions made by the model. In the confusion matrix, there are 45 false positives (anomaly predicted as normal) and 3987 true negatives (normal predicted as normal). There are also 73 false negatives (normal predicted as anomaly) and 3453 true positives (anomaly predicted as anomaly). In total, the model appears to have made $45 + 73 = 118$ incorrect predictions. Out of 3500 total predictions, that means the model has a success rate of $(3500 - 118)/3500 = 96.6\%$.

Here's a breakdown of the terms used in the confusion matrix:

```
1   import numpy as np
2   import pandas as pd
3   import matplotlib.pyplot as plt
4   from pandas.api.types import is_numeric_dtype
5   from sklearn.ensemble import RandomForestClassifier
6   import warnings
7   from sklearn import tree
8   from sklearn.model_selection import train_test_split
9   from sklearn.neighbors import KNeighborsClassifier
10  from sklearn.linear_model import LogisticRegression
11  from sklearn.preprocessing import StandardScaler, LabelEncoder
12  from sklearn.feature_selection import RFE
13  import itertools
14  import joblib
15
16  train=pd.read_csv('/Users/balajia/Desktop/CCN_project/Train_data.csv')
17  test=pd.read_csv('/Users/balajia/Desktop/CCN_project/Test_data.csv')
18
19  train.describe()
20  train.describe(include='object')
21  train.isnull().sum()
22  total = train.shape[0]
23  missing_columns = [col for col in train.columns if train[col].isnull().sum() > 0]
24  for col in missing_columns:
25      null_count = train[col].isnull().sum()
26      per = (null_count/total) * 100
27      print(f"{col}: {null_count} ({round(per, 3)}%)")

PROBLEMS   TERMINAL   OUTPUT   PORTS   DEBUG CONSOLE

/usr/local/bin/python3.11 /Users/balajia/Desktop/CCN_project/KNN_model.py
(base) balajia@BALAJIs-MacBook-Air CCN_project % /usr/local/bin/python3.11 /Users/balajia/Desktop/CCN_project/KNN_model.py
Class distribution Training set:
class
normal    13449
anomaly   11743
Name: count, dtype: int64
FrozenTrial(number=0, state=1, values=[0.9843874040751521], datetime_start=datetime.datetime(2024, 5, 17, 16, 6, 15, 297469), datetime_complete=datetim
e.datetime(2024, 5, 17, 16, 6, 15, 856427), params={'KNN_n_neighbors': 12}, user_attrs={}, system_attrs={}, intermediate_values={}, distributions={'KNN
_n_neighbors': IntDistribution(high=16, log=False, low=2, step=1)}, trial_id=0, value=None)
Train Score: 0.9854825904502665
Test Score: 0.9843874040751521
(base) balajia@BALAJIs-MacBook-Air CCN_project %
```

Figure 1.7 Accuracy of the k-nearest neighbours (k-NN) model.

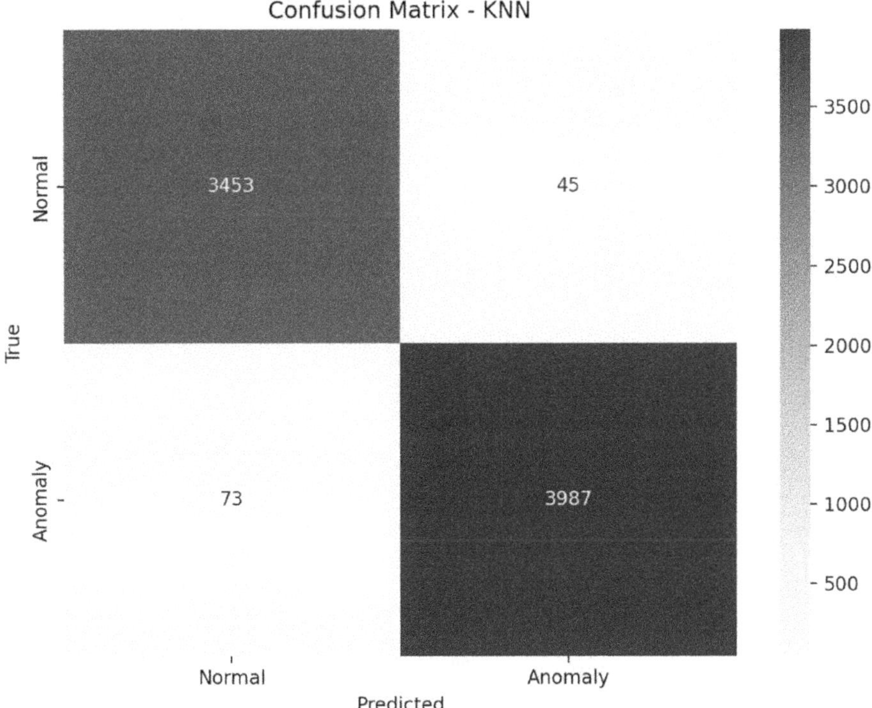

Figure 1.8 Confusion matrix.

- True Positive (TP): The model correctly predicts the positive class.
- True Negative (TN): The model correctly predicts the negative class.
- False Positive (FP): The model incorrectly predicts the positive class.
- False Negative (FN): The model incorrectly predicts the negative class.

Figure 1.9 shows the receiver operating characteristic (ROC) curve performance of the k-NN classification algorithm. ROC curve value shows the performance of the k-NN model for the access control mechanism and achieves a higher value between true positive and false positive rates.

Table 1.1 compares various machine learning models with the proposed model.

1.4.2 Limitation of the proposed work

- **Limited Tag Capacity:** The system may have a limit on the number of authorized RFID tags it can store, potentially restricting the number of users or access points. Managing a large database of tags and

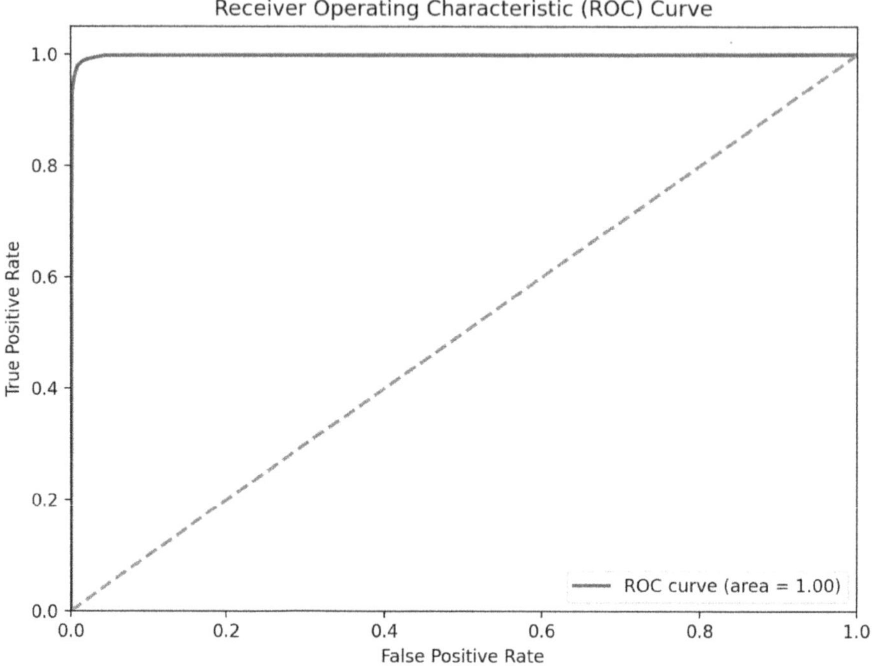

Figure 1.9 ROC curve.

Table 1.1 Comparison of Machine Learning Models

Authors	Model Name	Accuracy (%)
Hsu et al. [19]	SVM	92.3
Mishra [8]	CNN	94.4
Karimi [20]	PCA	90
Sarwar et al. [21]	ANN	95.74
	Proposed model	98.43

associated access patterns for AI-based anomaly detection might require additional memory and computational resources. This limitation can be addressed by periodically updating the system and optimizing memory usage.

- **Dependency on Power:** Continuous power is needed for the system to operate; a power outage could render the system non-functional temporarily. While backup power solutions can mitigate this risk, the AI anomaly detection system also depends on consistent power for real-time monitoring and analysis. Implementing uninterruptible power supplies and energy-efficient components can help reduce this dependency.

- **Vulnerability to Spoofing:** Despite security measures, the system may still be vulnerable to RFID spoofing attacks, where unauthorized individuals attempt to clone authorized RFID tags. Although AI-based anomaly detection can help identify unusual access patterns that might indicate spoofing, additional security measures such as encryption and multi-factor authentication may be necessary to fully address this risk. Implementing hardware-based security tokens and periodic key updates can further enhance protection.
- **Initial Cost:** The initial cost of implementing an RFID door lock system with AI-based anomaly detection can be higher than traditional locking mechanisms. However, the long-term benefits in terms of security, convenience, and scalability often outweigh the initial investment.

1.5 CONCLUSION AND FUTURE ENHANCEMENTS

This chapter presents the design and implementation of an RFID door lock system using Arduino and FreeRTOS, enhanced with AI-based anomaly detection using k-NN algorithm. The system integrates RFID technology for access control and servo motor control for the door locking mechanism. Through experimentation and testing, the feasibility and functionality of the proposed solution in providing secure, convenient, and intelligent access control have been demonstrated.

In conclusion, the RFID door lock system offers several advantages, including enhanced security through AI anomaly detection, convenience, and scalability. By leveraging RFID technology, FreeRTOS for multitasking, and AI for real-time monitoring and anomaly detection, a reliable and efficient solution suitablefor various applications, from residential to commercial settings, has been developed. This innovative approach not only improves security but also ensures adaptability to future advancements and diverse use cases. The integration of k-NN-based anomaly detection is a significant advancement in access control systems, as it allows for continuous monitoring and proactive identification of potential security threats. This capability enhances the overall security posture by enabling timely responses to suspicious activities, thereby reducing the risk of unauthorized access and potential breaches. Moreover, the flexibility of the system makes it an ideal solution for a wide range of environments. In residential settings, it provides homeowners with peace of mind through enhanced security and ease of use. In commercial and institutional contexts, it supports efficient access management and ensures that only authorized individuals can enter sensitive areas. The system's adaptability also means it can be customized to meet specific security requirements, making it a versatile tool in various industries.

The use of FreeRTOS ensures that the system can handle multiple tasks efficiently, maintaining high performance and reliability. This real-time operating system is crucial for managing the various components and processes involved in the RFID door lock system, ensuring that the system responds promptly to RFID tag detections and AI alerts. The scalability of the system means that it can be expanded or modified as needed without significant overhauls. This makes it a cost-effective solution in the long term, as it can grow and evolve with changing security needs and technological advancements.

1.6 FUTURE ENHANCEMENTS

Moving forward, there are numerous avenues for further enhancement and expansion of the RFID door lock system with AI-based anomaly detection. One key area for development involves the implementation of advanced security features to fortify the system against potential vulnerabilities and security threats. This includes exploring the integration of more robust encryption algorithms and tamper-resistant mechanisms to safeguard against unauthorized access attempts. Employing advanced cryptographic techniques will ensure that the communication between RFID tags and the control system remains secure and impervious to interception and cloning attacks. Efforts will be directed towards integrating the system with popular smart home platforms such as Amazon Alexa, Google Home, and Apple HomeKit to enable seamless interoperability and enhanced user control. This integration will allow users to manage access controls through voice commands and centralized smart home applications, thereby enhancing user convenience and control over their home security environment. Developing a dedicated mobile application will further empower users with intuitive remote access and management capabilities, fostering greater flexibility and convenience. This application can also leverage AI to provide real-time alerts and detailed insights into access patterns and potential security threats. The mobile app could include features such as remote locking and unlocking, access logs, and notifications of suspicious activities, allowing users to respond quickly to potential security incidents.

Optimizing energy consumption through the implementation of energy-efficient strategies will be essential to prolong the system's battery life, particularly in scenarios where continuous operation is required. The AI component can assist in optimizing energy usage by intelligently managing system resources based on usage patterns and environmental factors. For instance, the system could enter a low-power mode during periods of inactivity or utilize energy-harvesting technologies to supplement battery power. Expanding the system's capabilities to include biometric authentication methods, such as fingerprint or facial recognition, can provide multi-factor authentication for enhanced security. Combining RFID technology with biometrics will ensure that access is granted only when both an authorized RFID tag and a verified biometric credential are present.

Exploring the use of blockchain technology for secure and immutable logging of access events can further enhance the system's reliability and trustworthiness. Blockchain can provide a tamper-proof record of all access attempts, ensuring transparency and accountability in access control management. Integrating machine learning models for predictive analytics can help in identifying and mitigating potential security threats before they occur. By analysing historical access data, the system can learn to recognize patterns that precede security breaches, enabling pre-emptive actions to be taken.

The integration of enhanced real-time monitoring and alerting features will enable users to stay informed about access events and security incidents. By leveraging AI to analyse and interpret data, users can gain deeper insights into security trends and respond more effectively to potential threats, thereby enhancing overall situational awareness and responsiveness. Implementing a centralized dashboard for security monitoring can provide users with a comprehensive overview of their security status and enable quick decision-making. Ensuring compliance with industry standards and regulations will be critical for the widespread adoption of the RFID door lock system. Adhering to standards such as ISO/IEC 14443 for RFID and General Data Protection Regulation (GDPR) for data privacy will ensure that the system meets the necessary security and privacy requirements. Continuous improvement based on user feedback is essential. Regular updates and enhancements to the system based on user experiences and needs can help ensure that the system remains relevant and effective in addressing evolving security challenges.

Overall, the combination of RFID technology, FreeRTOS, and AI-based anomaly detection in this door lock system represents a significant step forward in access control solutions. It offers a sophisticated, reliable, and adaptable security system that can meet the demands of modern security challenges while providing a user-friendly experience. By continually enhancing security features, integrating with smart home ecosystems, optimizing energy efficiency, and expanding the system's capabilities, a comprehensive and robust access control solution can be developed to meet the evolving needs of modern security challenges.

REFERENCES

1. Verma, G. K., & Tripathi, P. (2010). "A Digital Security System with Door Lock System Using RFID Technology," *International Journal of Computer Applications*, Vol. 5, No. 11, pp. 6–7.
2. Komol, M. M. R., Podder, A. K., Ali, M. N., & Ansary, S. M. (2018). "RFID and Finger Print Based Dual Security System: A Robust Secured Control to Access through Door Lock Operation," *American Journal of Embedded Systems and Applications*, Vol. 6, No. 1, pp. 15–22.

3. Edozie, E., & Vilaka, K. (2020). "Design and Implementation of a Smart Sensor and RFID Door Lock Security System with Email Notification." *International Journal of Engineering and Information Systems*, Vol. 4, No. 7, pp. 25–28.

4. Soni, S., Soni, R., & Waoo, A. A. (2021). "RFID-Based Digital Door Locking System," *Indian Journal of Microprocessors and Microcontroller (IJMM)*, Vol. 1, No. 2, p. 17.

5. Nath, S., Banerjee, P., Biswas, R. N., Mitra, S. K., & Naskar, M. K. (2016). "Arduino Based Door Unlocking System with Real Time Control," In *2016 2nd International Conference on Contemporary Computing and Informatics (IC3I)*, IEEE, pp. 791–795. doi:10.1109/IC3I.2016.7917989

6. Baikerikar, J., Kavathekar, V., Ghavate, N., Sawant, R., & Madan, K. (2021). "Smart Door Locking Mechanism," In *2021 4th Biennial International Conference on Nascent Technologies in Engineering (ICNTE)*, NaviMumbai, India, pp. 1–4, doi:10.1109/ICNTE51185.2021.9487704

7. Venkatraman, S., Varshaa, R. R., & Vigneshwary, P. (2021). "IoT based Door Open or Close Monitoring for Home Security with Emergency Notification System Using LoRa Technology," In *2021 7th International Conference on Advanced Computing and Communication Systems (ICACCS)*, Coimbatore, India, pp. 173–178, doi:10.1109/ICACCS51430.2021.9441890

8. Mishra, R., Ransingh, A., Behera, M. K., & Chakravarty, S. (2020). "Convolutional Neural Network Based Smart Door Lock System," In *2020 IEEE India Council International Subsections Conference (INDISCON)*, Visakhapatnam, India, pp. 151–156, doi:10.1109/INDISCON50162.2020.00041

9. Saravanan, S. K., Asan Nainar, M., & Santhana Marichamy, V. (2019). "Android Based Smart Automation System Using Multiple Authentications," *IRE Journals*, Vol. 3, No. 6, pp. 59–65.

10. Park, Y. T., Sthapit, P., & Pyun, J. Y. (2009). "Smart Digital Door Lock for Home Automation." In *Presented at TENCON 2009-2009 IEEE Region 10 Conference*, Singapore, IEEE. doi:10.1109/TENCON.2009.5396038

11. Gindi, S., Shaikh, N., Beig, K., & Sabuwala, A. (2020). "Smart Lock System Using RFID," *International Research Journal of Engineering and Technology (IRJET)*, Vol. 7, No. 7, p. 2853.

12. Mishra, S., Mohite, O., & Kharat, S. (2022). "Smart Door Lock System Using Arduino." *International Research Journal of Modernization in Engineering Technology and Science*, Vol. 4, No. 4, p. 1503.

13. Qureshi, S., Sharma, Y., Wadwekar, S., Sharma, Y., Panwar, Y.S., Sankhala, R., & Gupta, P. K. (2020). "Face Recognition (Image Processing) based Door Lock using OpenCV, Python and Arduino." *International Journal for Research in Applied Science & Engineering Technology (IJRASET)*, Vol. 8, No. VI, pp. 1208–1214.

14. Al Bakry, A. M., & Resan, R. D. (2016). "Smart Phone-Arduino Based Smart Door Lock/Unlock Using RC4 Stream Cipher Implemented in Smart Home." *International Journal of Advanced Computer Technology (IJACT)*, Vol. 5, No. 5, pp. 14–18.

15. Kumbhar, N. N., & Mane Deshmukh, P. V. (2017). "Smart Door Locking System using Wireless Communication Technology." *International Journal for Research in Applied Science & Engineering Technology (IJRASET)*, Vol. 5, No. VIII, p. 507.

16. Shanthini, M., Vidya, G., & Arun, R. (2020). "IoT Enhanced Smart Door Locking System," In *Presented at the 2020 Third International Conference on Smart Systems and Inventive Technology (ICSSIT)*, IEEE. doi:10.1109/ICSSIT48917.2020.9214288

17. Sahu, S., Singh, R., Arya, P., & Nirala, R. (2022). "Smart Home Automation Lighting System and Smart Door Lock using Internet of Things." In *Presented at the 2022 4th International Conference on Advances in Computing, Communication Control and Networking (ICAC3N)*, IEEE. doi:10.1109/ICAC3N56670.2022.10074243

18. Ravi Guntur. API security: Access behavior anomaly dataset. Available at: https://www.kaggle.com/datasets/tangodelta/api-access-behaviour-anomaly-dataset (accessed on 18 January 2021).

19. Hsu, H.-T., Jong, G-J., Chen, J.-H., & Jhe, C.-G. (2019). "Improve Iot Security System of Smart-Home By Using Support Vector Machine," In *2019 IEEE 4th International Conference on Computer and Communication Systems (ICCCS)*, Singapore, pp. 674–677, doi:10.1109/CCOMS.2019.8821678

20. Karimi, K., Kabrane, M., Hassan, O., Badouch, A., & Krit, S. (2020). "Secure Smart Door Lock System Based on Arduino and Smartphone App," *Journal of Advanced Research in Dynamical and Control Systems*, Vol. 12, pp. 407–414.

21. Sarwar, N., Bajwa, I. S., Hussain, M. Z., Ibrahim, M., & Saleem, K. (2023). "IoT Network Anomaly Detection in Smart Homes Using Machine Learning," *IEEE Access*, Vol. 11, pp. 119462–119480, doi:10.1109/ACCESS.2023.3325929.

## Chapter 2

Issues with privacy and cybersecurity in the multi-cloud environment in smart cities

Ramiz Salama and Fadi Al-Turjman

2.1 INTRODUCTION

The usage of cloud computing technology is seen in business growth initiatives. You can select from a range of cloud service providers, such as IaaS, SaaS, and PaaS, based on your requirements. IaaS (Infrastructure as a Service) provides virtualized computing resources such as servers, storage, and networks over the internet. PaaS (Platform as a Service) offers a development platform and environment for building, deploying, and managing applications without handling the underlying infrastructure. SaaS (Software as a Service) delivers fully functional software applications accessible via web browsers on a subscription basis. Users are free to use any service here for any reason at all. At first, cloud computing was integrated with academics and business to establish a hub for innovation. The process of providing the required computing and storage resources has become simpler thanks to the recently built enterprise representation system. An organization's financial outcomes are directly impacted by the value it sets on its data. Two qualities that constitute data are legitimacy and consistency, and organizations must protect and segregate their data. Many modern enterprises choose to employ numerous cloud platforms because cloud computing provides greater flexibility and access to a wider choice of services. On the contrary, a lot of private and public cloud networks function more like closed systems than like open ones. Businesses need to solve the challenges posed by insufficient interconnectivity between cloud networks to fully profit from multi-cloud computing. To keep utilizing the same apps once data and apps have been transferred to this multi-cloud architecture, users must be safeguarded. Many different service providers maintain information about consumer accounts in the cloud. It is vital for customers who use several clouds to maintain their credentials across those services. But this also means that there are more duplicate user data, which is dangerous for cloud service providers and their customers. Accurately identifying data and selecting suitable cloud services are the first steps in an organization's data migration into a multi-cloud architecture. Many businesses are beginning to worry about the security of sensitive data and important apps kept

DOI: 10.1201/9781032618173-3

on the cloud because of this distributed structure. Another growing concern is how to recover any erased material. The past several decades have seen an increase in computational power and the amount of information available over the Internet, leading to the development of a complex architecture known as multi-cloud, which uses fog and heterogeneous cloud computing. Multi-cloud architecture encompasses a range of techniques and services from several providers that use different cloud infrastructures and deployment patterns. Cloud customers can process and scale on-demand services from several environments thanks to multi-cloud. Multi-cloud's wide range of controls ensures transparency when it comes to Internet-based access control and security for data or service access. A multi-cloud deployment strategy that makes use of transparent access control techniques for hosted services and apps can help reduce cybersecurity threats in addition to lowering energy requirements for cloud service providers. Furthermore, multi-cloud offers an efficient means of avoiding the hardening of disparate cloud services by offering necessary functionality under a centralized architecture [1–5]. Although a specific demand may need the allocation of both virtual and physical resources, the multi-cloud architecture's robustness and decentralization raise concerns about the safety implications of resource distribution. We offer a fresh perspective on how to handle these issues related to data security and access control through our research.

2.2 RELATED WORK

Pan et al. proposed the usage of automatic marks from private clouds. In the field of cybersecurity, digital signatures are gaining popularity; yet, verifying signatures takes longer than producing recall-collated values. Employing Guess as a universal server for creating elliptic curve (EC) structures with the algorithm running on it, this inquiry report suggests employing the ECDSA-based EC digital signature approach with a key length of 256 bits. Guess makes use of threads to increase speed and processing capacity; however, many defense experts rely on subcontracting to produce and authenticate signatures. Guess makes use of additional software components for simple scaling and upgrades. Data presented in the research lend credence to the idea that Guesses can be used as proof of secure network transactions, adaptable characteristics, and optimization constraints. An increasing trend in the adoption of cloud-based multimedia apps is reported by Yang et al. In spite of potential security risks, we employed a cluster of video data in the context of cloud computing in this instance to gather additional knowledge over time. Utilizing cloud computing's benefits to keep data in an organized manner and avoid overpaying is the best way to store this kind of information. Every video clip is encrypted for each user individually, and recommendations are provided to improve the effectiveness and speed of customer feature updates, such as giving up exclusive features or getting rid of outdated ones that can be improved later. Signing new features and rescinding old ones are examples of dynamically changing user options. Respecting Cheng et al. introduce an identity-based strategy that provides a clear justification for ensuring accountability to address the maintenance of acceptability and privacy in cloud computing. To maintain accountability and audits, the authors provide a method for safeguarding cloud computing against adversarial attacks that compares data and outcomes using identity-based encryption [6–10]. Using an SSL-based encryption protocol, Subramaniam Jegadeesan et al. have suggested utilizing several forms of mutual authentication inside their framework to protect mobile consumers and service providers from any potential security issues. The architecture that was shown offered a helpful method of measuring and controlling processing costs in addition to enabling session key changes from either side.

2.3 METHODOLOGY

System model Figure 2.1 depicts the multi-cloud data storage system with three distinct entities.

The following list contains three distinct network entities:

1. **Users (US):** People who must keep a substantial amount of data files on many cloud providers.

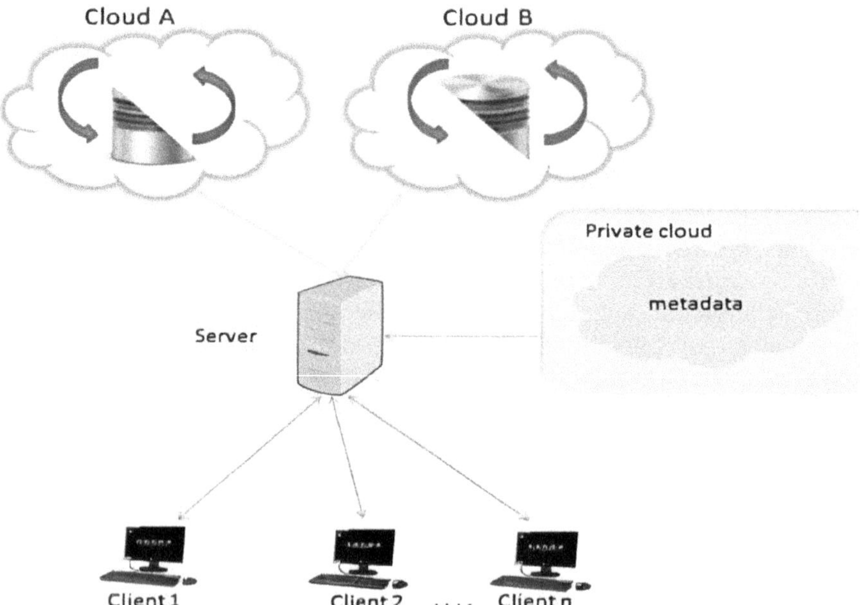

Figure 2.1 System model.

2. A number of cloud storage servers (CSSs) from different cloud service providers, each with a sizable amount of processing and storage capacity.
3. A third-party auditor (TPA) who attests to the veracity of the information provided by cloud users.

Each CSS stores and maintains a different amount of the client's data in a distributed storage server setup, and a coordinator (O) oversees all interactions between TPA and a few CSSs. In the United States, TPA and various CSSs are regarded as semi-confided substances; associations and TPAs cannot be reconciled when TPAs and CSSs are separated. It's critical to strike a balance between ensuring the accuracy of data stored in cloud storage and protecting data privacy. Asking their cloud service provider (CSP) for cloud file storage is the first step a client takes when seeking for it; however, if the file is blank, they next focus on finding out about access and space on the cloud server [11,12]. Users will have the option to decide whether the data they have contributed needs more protection or care after completing the previous steps. The file will be divided into many virtual machines (VMs) if sensitivity is an issue; if not, it will be housed in a single VM. Our technology makes sure that owned data stored in the cloud is automatically encrypted and that the encryption process is maintained—a prerequisite for restitution. At the moment, the supplied data is encrypted with the ECIES technique. The recommended approach is shown in Figure 2.2 as a general block diagram; further information is provided in a supplement. The recommended approach is put into practice as follows:

Step 1: Users must enter a unique username and password combination together with their email address to register on this website. Their submitted password serves as the encryption function's private key or personal key. If you do not have the same private key that was used to encrypt this file, you will not be able to decode it. The use of both of these keys—public and private—is necessary for this file.

Step 2: Elliptic Curve encryption relies on a pair of matching keys: a public key for encrypting data and a private key for decrypting it. The Elliptic Curve Integrated Encryption Scheme (ECIES) is a robust public-key-based solution that not only facilitates secure key exchange but also provides authenticated encryption, ensuring both confidentiality and integrity of the transmitted data. Public-key methods are a crucial tool for safe information sharing, notwithstanding their potential challenges in development and implementation. By using advanced methods for key generation and secure transmission over a cloud server, we may reduce network traffic and improve the security of our authentication system [13–15]. ECIES with elliptical curves can be used as a backup method when private keys are not needed for data encryption or signing using our public key.

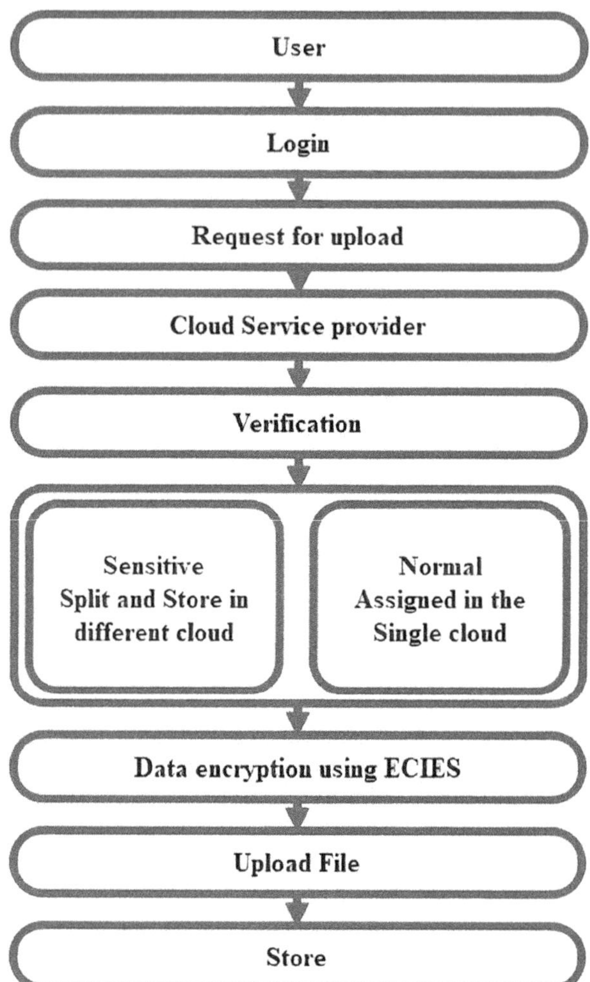

Figure 2.2 Cloud computing cryptography-based data storage.

Step 3: Check of Record: The file's status is then determined.
- **Normal File**: Designating the file as non-sensitive ensures that only one VM will handle it.
- **Sensitive File**: When dealing with a sensitive file's data, its division among smaller portions is determined by its overall size and randomly assigned VMs are used for each segment of the information. The number of files that our VMs can process simultaneously is unlimited.

Step 4: Data Encryption using ECIES algorithm: User data and signatures are securely stored using the Secure Hash technique, and data packets sent by mobile nodes are encrypted before being transmitted to the cloud server by means of the ECIES technology. With ECIES, you may create digital signatures and perform encryption and other operations while preserving authentication security. The EC uses smaller-sized keys for generation and exchange, which lowers the bandwidth required to transport keys across networks connected by cloud servers. This chapter examines the fundamentals of EC cryptography, which involves selecting a curve that is frequently used for key exchange when signing public documents and demonstrating that it can also be used to encrypt data via ECIES. We produce our private key by choosing a random number and using it to represent the gradient of the related line, starting with the public key of an x–y point. In the block chain implementation, public-key pairs will be created using ECs. An ECC technique is more secure than Rivest–Shamir–Adleman's (RSA's) 3072-bit strategy, utilizing 256-bit public keys. The EC provides similar security with fewer key sizes than the RSA. Smaller keys and quicker processing are the two main benefits of using EC-based cryptography security.

Developers of cloud-based solutions usually think that hosting companies provide enough security, but unforeseen acts by the cloud operator can provide a lot of risks. Even if encryption might not be able to stop all attacks on data, enabling harmful activity might cause massive data leaks. After carefully analyzing typical cloud operations, we built two threat models, based on our discovery that cloud operators are the primary source of danger.

Workers on the cloud side are likely to wish to get around security protocols and get the data that this strategy takes into account. Workers who use the cloud have access to the server and the encryption key.

We'll assume for this scenario that the cloud-side operators want to access the data and information without authorization. The cloud administrator is aware of the information stored on the cloud server. The high level of encryption does not prevent operators from decrypting the data and determining what is included.

Definition 1 can be used to construct these two threat models.

Definition 1: $K \to D$, the crucial K needed to decode a cloud-based information package. Presume cloud administrators use K to D without permission from the data owner.

To ensure that clients can effectively verify the foundational aspects of information security, our proposed architecture must simultaneously achieve the following specific objectives:

To ensure that clients can effectively verify the foundational aspects of information security, our proposed architecture distributes data across multiple cloud servers, we aim to achieve a higher level of data security and capacity. This approach is designed to protect data from external threats, such as those posed by terrorists or criminals. Additionally, all data must be encrypted during transmission to ensure its confidentiality and integrity. Our strategy also emphasizes extremely high data processing efficiency. Furthermore, by minimizing computational and communication costs, our approach reduces operational overhead and enhances overall system performance.

2.4 RESULTS AND DISCUSSION

The Java 8 functionality is available in the most recent version of the NetBeans IDE; cloud sim was utilized to carry out the experiment. Our workstation, which had an Intel Core i5 processor and 8 GB RAM, was used to upload data using a Java-based graphical user interface that was integrated with cloud sim technology. The Cloud Analyst program was also utilized to locate data centers. The final version can only be run with a basic configuration; nevertheless, the results may not be worth the effort, and the duration may differ depending on how effective it is. The encryption/decryption times of the suggested model are compared to well-known cryptographic techniques such as AES/ECC in Table 2.1.

As file sizes increase, file decryption takes longer. Processing files with a total size of 10,240 bytes or less is observed to be completed at a rate of

Table 2.1 Encryption and Decryption Time Taken

Technique	File Size (KB)	Encryption Time (ms)	Decryption Time (ms)
AES	10	645	578
	20	815	632
	30	1187	845
	40	1390	1011
ECC	10	617	543
	20	789	614
	30	1034	866
	40	1345	987
Proposed ECIES	10	563	395
	20	715	96
	30	893	632
	40	1201	846

approximately 578 ms when using AES encryption algorithms on the data; this is completed in approximately 542.7 ms when using an ECC-based encryption technique; and, finally, with our special method developed here, we were able to complete the operation very quickly in only an approximate value of somewhere between 394 and 5 ms. Several analysis techniques were used when the same test was run on a file containing data up to the maximum limit of (file size = 30 KB). Technique A took between 845 and 866 ms, while Technique B was slower. Interestingly, our newly developed method, the proposed model, resulted in a far lower millisecond count, making it far more effective than the rest. The proposed ECIES model, when compared to the existing ECC and AES models, shows that it has less decryption time even when dealing with larger files. We found an intriguing trend in this study: all models demonstrated faster file decryption times than their corresponding encryption times. This can be explained by the fact that additional steps, like key generation, are needed during file encryption and that there are no extra overheads when decrypting files using similar modes. And a comparison of the uploading and downloading times for the model is shown in Table 2.2.

When using AES and ECC techniques for encryption and decryption, uploading a data set that includes at least one file of exactly 30 kilobytes (KB) takes longer compared to our proposed model. Through testing, we discover that when these files are encrypted using two different methods, namely, the proposed ERIS method alone and both AES/ECC together, the latter takes approximately 5600 ms, whereas the former takes less than half

Table 2.2 Time Taken for Downloading and Uploading Time Based on a File Size

Technique	File Size (KB)	Uploading Time (ms)	Downloading Time (ms)
AES	10	3852	2984
	20	4671	3756
	30	5453	4290
	40	6814	5101
ECC	10	3917	2762
	20	4567	3841
	30	5740	4289
	40	6853	5574
Proposed ECIES	10	2145	1465
	20	3365	2141
	30	4211	2965
	40	5236	3954

that time, taking just over 4210 ms. Additionally, when processing small file sizes (less than or equal to 10 KB), the suggested model took somewhat longer (approximately 2154 ms) than the AES and ECC approaches (around 3890 ms). Downloading took less time than uploading because the suggested model included AES and ECC methods. With a document of approximately 20 KB or less, computations show that to reach comparable results, traditional AES and ECC techniques need to be applied almost close to/halfway up toward/almost reaching up to/close-by/more or less/around LY/toward –3709 ms, while our genuinely proposed ECIES model needs to be applied almost no more than/approximately/barely over/about/nearly/exactly/scarcely exce

eding/ranging from –2294 ms. AES and ECC techniques required more than 5400 ms for high file volumes. The investigation indicates that the ECIES technique yields quicker download speeds than the other methods. Consequently, the proposed model could process the data in less than 3954 ms.

2.5 CONCLUSION

We created an intelligence algorithm called ECIES to accomplish our goal of prohibiting cloud service providers from accessing users' personal data. Files can be labeled as sensitive or non-sensitive by users. The ECIES algorithm is necessary to secure critical files on VMs by dividing them into smaller, more manageable chunks. The usefulness of using this information to track down the people who tampered with the evidence cannot be overstated, and as long as user privacy is upheld, it is possible to improve the trustworthiness of the data. The algorithm in cloud sim is implemented using the Java platform. Experimental evaluations have demonstrated that our suggested approach is a very strong defense against significant cloud-side threats. According to the results of the experiment, the newly suggested scheme performs better than the existing one in terms of the encryption and decryption times of the data handled by both systems. Increasing the availability of future retrievals has made protecting the retrieved data from duplicate entries a critical priority.

REFERENCES

1. Tabassum, N., Naeem, H., & Batool, A. (2023). The data security and multi-cloud privacy concerns. *International Journal for Electronic Crime Investigation*, 7(1), 49–58.
2. Rao, P. M., & Deebak, B. D. (2023). Security and privacy issues in smart cities/industries: Technologies, applications, and challenges. *Journal of Ambient Intelligence and Humanized Computing*, 14(8), 10517–10553.
3. Stergiou, C. L., Bompoli, E., & Psannis, K. E. (2023). Security and privacy issues in IoTbased big data cloud systems in a digital twin scenario. *Applied Sciences*, 13(2), 758.

4. Iftikhar, A., Qureshi, K. N., Shiraz, M., & Albahli, S. (2023). Security, trust and privacy risks, responses, and solutions for high-speed smart cities networks: A systematic literature review. *Journal of King Saud University-Computer and Information Sciences*, 35(9), 101788.

5. Singh, T., & Kumar, A. (2023, February). Analyzing security and privacy issues for multi-cloud service providers using nessus. In *2023 Fifth International Conference on Electrical, Computer and Communication Technologies (ICECCT)* (pp. 01–08). Erode, India: IEEE.

6. Mostafa, N., Kotb, Y., Al-Arnaout, Z., Alabed, S., & Shdefat, A. Y. (2023). Replicating file segments between multi-cloud nodes in a smart city: A machine learning approach. *Sensors*, 23(10), 4639.

7. Hashem, I. A. T., Usmani, R. S. A., Almutairi, M. S., Ibrahim, A. O., Zakari, A., Alotaibi, F., ... & Chiroma, H. (2023). Urban computing for sustainable smart cities: Recent advances, taxonomy, and open research challenges. *Sustainability*, 15(5), 3916.

8. Pandey, R., Rajput, I. S., Tyagi, S., Koranga, M., & Sharma, P. K. (2024). Enhancing smart city resilience: Integrating cloud computing and 5g for secure and sustainable urban environments. In *Secure and Intelligent IoT-Enabled Smart Cities* (pp. 77–90). Pennsylvania, USA: IGI Global.

9. Verma, R., & Jailia, M. (2023). A comprehensive taxonomy and systematic review of intelligent attack defense systems for multi-cloud environment. *International Journal of Intelligent Systems and Applications in Engineering*, 11(9s), 444–454.

10. Tricomi, G., Giacobbe, M., Ficili, I., Peditto, N., & Puliafito, A. (2024). Smart city as cooperating smart areas: On the way of symbiotic cyber–physical systems environment. *Sensors*, 24(10), 3108.

11. Nizamudeen, S. M. T. (2023). Intelligent intrusion detection framework for multi-clouds–IoT environment using swarm-based deep learning classifier. *Journal of Cloud Computing*, 12(1), 134.

12. Nizamudeen, S. M. T. (2023). Intelligent intrusion detection framework for multi-clouds–IoT environment using swarm-based deep learning classifier. *Journal of Cloud Computing*, 12(1), 134.

13. Mostafa, A. M., Ezz, M., Elbashir, M. K., Alruily, M., Hamouda, E., Alsarhani, M., & Said, W. (2023). Strengthening cloud security: An innovative multi-factor multi-layer authentication framework for cloud user authentication. *Applied Sciences*, 13(19), 10871.

14. Surianarayanan, C., & Chelliah, P. R. (2023). Integration of the internet of things and cloud: Security challenges and solutions–a review. *International Journal of Cloud Applications and Computing (IJCAC)*, 13(1), 1–30.

15. Stroupe Jr, T. R. (2023). A Qualitative Comparative Analysis of Data Breaches at Companies with Air-Gap Cloud Security and Multi-Cloud Environments (Doctoral dissertation, Purdue University).

## Chapter 3

Blockchain technology in healthcare systems along with cybersecurity

Ramiz Salama and Fadi Al-Turjman

3.1 INTRODUCTION

The application of blockchain technology is attracting the interest of many academics, institutions, and businesses, particularly in relation to the use of the virtual currency Bitcoin. A blockchain is a decentralized ledger that safely records transactions on a peer-to-peer network. It also simplifies and clarifies communication. Blockchain technology's primary objective is to allow two parties to safely manage transactions without the need for outside intervention [1–8]. Aside from blockchain, other cutting-edge advancements like artificial intelligence (AI), the Internet of Things (IoT), computer-generated reality, and man-made consciousness have revolutionized a variety of fields and sectors, including banking, protection, medical services, aviation, design, automotive, and electronics [9].

The application of blockchain technology in the medical services industry has improved patient–provider communication simplicity and transparency. Due to duplicates, the usage of various names and identities and their availability across multiple institutions, medical records are growing larger and more complex, but they haven't yet been simplified. Additionally, the importance of healthcare security has grown in terms of data security and deterring criminal behavior. Anybody with access to the data can examine patient information if unauthorized parties obtain it, making it possible for it to be used or sold. Maintaining the privacy of patient information is crucial to the effective management of healthcare. Industry 4.0 and blockchain technology can be used to address these problems by safeguarding data integrity and preventing manipulation and failure at any one time (Figure 3.1).

An open, decentralized digital ledger known as a blockchain keeps track of transactions across multiple computers in a manner that prevents the modification of one entry without also altering entries from subsequent blocks. A lengthy chain is created by blockchain technology and is linked to and verified by the initial "block." The record is ultimately referred to as blockchain. Due to the open auditing and public publication of every transaction, blockchain offers several accountabilities. Once input is made, none of the data posted to the blockchain may be changed. It illustrates

34

DOI: 10.1201/9781032618173-4

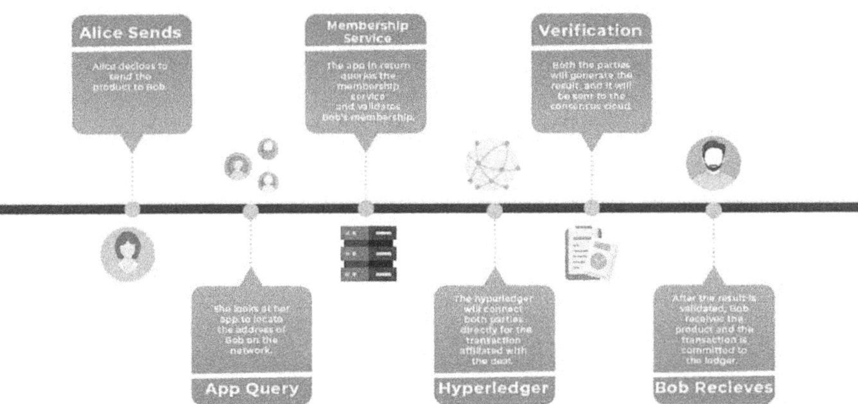

Figure 3.1 Hyperledger.

how reliable and consistent the data is. How fragile and hacker-friendly blockchain technology is is demonstrated by the fact that data is kept across networks rather than in a single database. In this sense, blockchain offers a great platform for creating new business models and competing with well-established firms [1–3]. Blockchain technology facilitates improved tracking of medical product sales for marketers. It will enable the pharmaceutical and healthcare industries to avoid creating fictitious prescriptions, aiding more patients to take their medications as directed. It assists in identifying the main source of falsification. The classification of patient records, the real-time preservation of medical history, and the immutability of these records are all guaranteed by blockchain technology. With its traditional gear, the hospital makes advantage of this decentralized network. Using the resources these devices store, scientists are able to record gauges for the treatments, medications, and fixes for a wide range of illnesses and problems [4,5].

3.2 RELATED WORK

A blockchain is an ordered set of linked blocks used for consistent data storage. Blocks are linked together based on their inputs, and even small tampering with the records within these blocks can break the bonds holding the blocks together. Because blockchain is replicated throughout a network of nodes, each of which has a stake in the network, it is frequently referred to as state machine replication. Permissioned and unpermitted blockchains are the two basic subcategories [4]. Transactions in a permissionless blockchain are verified by the entire public, but those in a permissioned blockchain are verified by certain organizations. Despite being more centralized, permissioned systems are faster and more scalable. On the contrary, permissionless blockchain systems are accessible to everybody. Data on blockchain technology is unchangeable due to its nature. A current product on the market for electronic medical record (EMR) monitoring is described by the authors in Ref. [5]. The writers of Ref. [6] provide additional details about the present status of blockchain technology research. The findings' numbers show how many publications focus on various blockchain application sectors. Less than 20% of agreements with other blockchain apps and more than 80% of documents, including clever agreements and authorization, are based on the Bitcoin basis. Most research attempts to draw attention to blockchain systems' inefficiencies as well as privacy and security flaws. However, there isn't a clear image or any hard evidence to support their claims. The authors examine a few of the most important blockchain protocols in Ref. [7]. The creators of blockchain protocols proposed the Hyperledger blockchain texture in Ref. [8], a distributed open-source system with pluggable features because of its architecture. By offering a publicly available platform to track transmitted data, it propels blockchain innovation. This Linux-based platform could change the way the sector functions. Actually, a large number of blockchain-related products currently available on the market make use of Hyperledger. The authors of Ref. [9] have talked about the various use cases and problems that blockchain technology and IoT may solve together. The research places the most emphasis on the applications of the shared digital economy. Furthermore, there are several well-founded models accessible for the engineering of IoT and blockchain. In Ref. [10], the writers proposed a capture method. Haze figures are used in medical services data utilizing a patient-driven methodology. The work recommended in this post can help collect patient records, which can then be input into our blockchain system of choice. Further engineering work is put into refining the haze computation by the authors of Ref. [11] to handle astute eHealth services more effectively. The authors of Ref. [12] delve into the complexities, background, and development of blockchain technology, as well as the revolutionary effects it has had on the IT and non-IT sectors. By carefully examining the literature, the writers of Ref. [13] have successfully conveyed the concept and importance of blockchain technology. With several examples, the authors of Ref. [14] provide a more thorough theoretical explanation of blockchain technology. The authors of Ref. [15] talked about how the banking industry has improved security and privacy. Using blockchain technology can assist in managing the complex issues related to autonomous, permissionless, decentralized systems. Additionally, it can offer some dependable methods for integrating security components within the banking industry (Figure 3.2).

Search phrases such as "healthcare systems (s)," "healthcare records (s)," "health care & system & records," and "healthcare blockchain" were used to discover relevant papers from a variety of sources, including ProQuest and Google Scholar databases, for the study on electronic health record

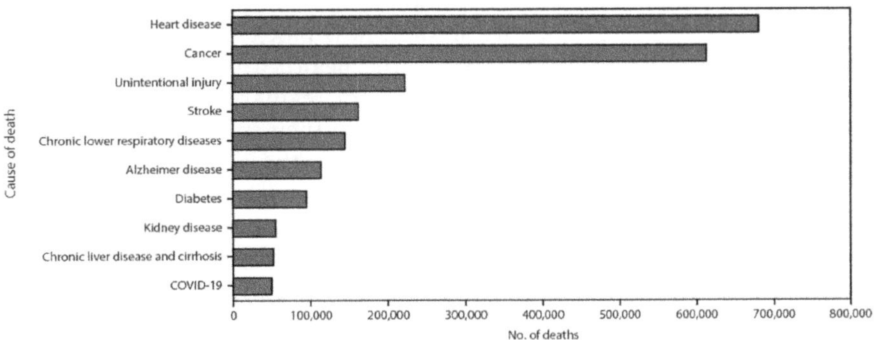

Figure 3.2 Disease link death rate in the USA.

(EHR) literature review. By restricting the citation of crucial research studies, this publication offers a succinct summary [8,11]. People have to provide their permission for medical records to be kept on file even as technology develops. Surprisingly, 87% of US individuals routinely get hard copies of these records, with over half getting them from medical professionals [5]. However, users find it difficult to provide information in EHR systems due to various challenges, including security. Rezaeibagha et al. considered the consequences for EHR systems with regard to security and privacy. Information health security and privacy were found to be significantly impacted by information sharing and integration. A recent study by Afrizal et al. looked at the efficiency of EHR systems from both an organizational and a personal perspective. Their research uncovered organizational boundaries and also brought to light deficiencies in senior management, a lack of teamwork, and a paucity of skilled labor. Personal limitations included not knowing how to use new software and having limited access to computers.

Blockchain provides a number of ways to lower barriers in EHR systems, and new technologies play a critical role in removing these kinds of obstacles. With the use of blockchain technology, it is now feasible to ensure that each transaction that occurs within a network is accurately recorded and permanently saved. Furthermore, no one individual is in charge of overseeing the computing operations necessary to carry out multi-computer transactional processes because the system is entirely distributed. The application of blockchain technology holds the potential to enhance the sustainable development objectives of the United Nations, particularly within the healthcare sector. Blockchain technology has the potential to make public sector services like EHRs more modern. Blockchain was looked into by Zhang and his colleagues as a possible way to safeguard patient data in healthcare systems. Secure data exchange settings show that the needs of patients come first. Using blockchain technology, which allows for the improved tracking of opioid prescriptions and more patient access to cancer-related patient data as well as other health services like telemedicine and insurance claims, is one way to enhance the administration of healthcare data. Using patient health records as an example, we showed how blockchain is changing the way that information is exchanged in the healthcare industry. A blockchain is a collection of records consisting of regularly organized blocks. Blocks are connected via links. Depending on their inputs and by altering the records in these blocks, the linkages between the blocks can be severed. Because blockchain is repeated across a network of nodes, each of which has a stake in the network, it is also known as state machine replication. There are two main types of blockchains based on whether they are permissioned or not: permissionless/permissioned blockchain [4]. Transactions in a permissionless blockchain are verified by the entire public, but those in a permissioned blockchain are verified by certain organizations. Despite being more centralized, permissioned systems are faster and more scalable. On the contrary, permissionless blockchain systems are accessible to everybody. Data on blockchain technology is unchangeable due to its nature.

A current product on the market for EMR monitoring is described by the authors in Ref. [5]. The writers of Ref. [6] provide additional details about the present status of blockchain technology research. The findings' numbers show how many publications focus on various blockchain application sectors. Less than 20% of agreements with other blockchain apps and more than 80% of documents, including clever agreements and authorization, are based on the Bitcoin basis. Most research attempts to draw attention to blockchain systems' inefficiencies as well as privacy and security flaws. However, there isn't a clear image or any hard evidence to support their claims. The authors examine a few of the most important blockchain protocols in Ref. [7]. Its creators proposed the Hyperledger blockchain texture in Ref. [8], a distributed open-source system with pluggable features because of its architecture. By offering a publicly available platform to track transmitted data, it propels blockchain innovation. This Linux-based platform could change the way the sector functions. Actually, a large number of the blockchain-related products currently available on the market make use of Hyperledger.

The research places the most emphasis on the applications of the shared digital economy. Furthermore, there are several well-founded models accessible for the engineering of IoT and blockchain. In Ref. [10], the writers proposed a capture method. Haze figures are used in medical services data utilizing a patient-driven methodology. The work recommended in this post can help collect patient records, which can then be input into our blockchain system of choice. Further engineering work is put into refining the haze computation by the authors of Ref. [11] to handle astute eHealth services more effectively. The authors of Ref. [12] delve into the complexities, background, and development of blockchain technology, as well as the revolutionary effects it has had on the IT and non-IT sectors. By carefully examining the literature, the writers of Ref. [13] have successfully conveyed the concept and importance of blockchain technology. With several examples, the authors of Ref. [14] provide a more thorough theoretical explanation of blockchain technology. The authors of Ref. [15] talked on how the banking industry has improved security and privacy. Using blockchain technology can assist in managing the complex issues related to autonomous, permissionless, decentralized systems. Additionally, it can offer some dependable methods for integrating security components within the banking industry. Zhang et al. have praised blockchain architecture for patient records as the best way to handle health records; however, there aren't many studies on the topic.

Some of the researches focussed on developing a blockchain-based management information system for EHRs in response to privacy and security concerns. The ledger database committer, orderer, endorser, and client are the six components that form the foundation of their design. The Fan-led group, however, paid little attention to the issues around personal data or the concept of digital currency. They left these subjects for further research to bolster the work of Fan et al. Griggs and colleagues sought to solve security problems that arose when utilizing blockchain by building a private network. Block can document J and K's past and current circumstances because of its robustness as a record-keeping method. RTT transactions come in two flavors: private and public. Private blockchains may be a useful tool for addressing privacy issues with the processing of personal data in the healthcare sector, according to research by Griggs and other experts. Privacy concerns could have an impact on people's decision to regularly use EHR systems. In their study, Sharma and co-authors used a soft systems approach to present qualitative evidence that adopting blockchain to share EHRs raises patient engagement opt-in percentages. They concentrated on the precision health care (PHC) initiative, which assembles discrete EHRs to promote public health and facilitating universal access. It has been demonstrated that the blockchain-based system concept improves collaboration through better access to patient records and boosts reliability in unreliable PHC systems. Researchers Esmaeilzadeh and Mirzaei examined the potential impact of blockchain on Health Information Exchange (HIE), and their conclusions suggest that the privacy-preserving characteristics of a blockchain-based system would be the primary reason for consumer preference. To simplify blockchain integration into EHR, Shahnaz and colleagues offered a way to allay worries about adaptability when introducing blockchain through recommended structural modifications. Future research may look into the advantages and disadvantages of applying blockchain technology in the medical industry. This is the most significant study on the impact of blockchain technology on patients' plans to use mediation to exchange their healthcare data to date. The application of blockchain technology in healthcare is still unknown, despite a number of recent research looking into how this technology might enhance health information management. Because there hasn't been much research on these subjects, we don't fully understand how extrinsic motivations and security perceptions affect the information systems utilized by healthcare practitioners.

3.3 METHODOLOGY

This study aims to assess the potential benefits and challenges of implementing blockchain technology in the healthcare industry. The methodology that was employed to conduct the investigation is covered in this section of the report. To conduct this evaluation, we followed four main procedures that included extracting and pre-processing the dataset, inspecting it, and interpreting it. If you want to know how blockchain technology and the healthcare sector interacted between 2016 and 2020, as shown by their indexation in WoS and Scopus, go no further than this dataset. This study conducted a bibliometric analysis of blockchain technology in the healthcare industry sector using the open-source statistics program R. The R desktop system now has the program installed and running. Many academic fields have investigated this bibliometrics method (Figures 3.3 and 3.4).

To find articles that included the term "blockchain in medical services" in their titles, watchwords, or altered compositions, this work conducted a literary survey. Finding blockchain applications in healthcare administration, or locations where its application has been proposed, was the primary objective of the review of previous studies. A range of online resources, such as ResearchGate and Google Scholar, in addition to the EBSCO database, the Web of Science, and the Applied Science and Technology Source, were used to choose the papers The advantages and disadvantages of blockchain-based technology in the healthcare industry were analyzed using 40 of the most recent manuscripts from chosen papers published between 2016 and 2020. Only written works published in the English language were taken into account, with a bias toward health-related periodicals. The poll's results provide a thorough overview of blockchain's potential applications

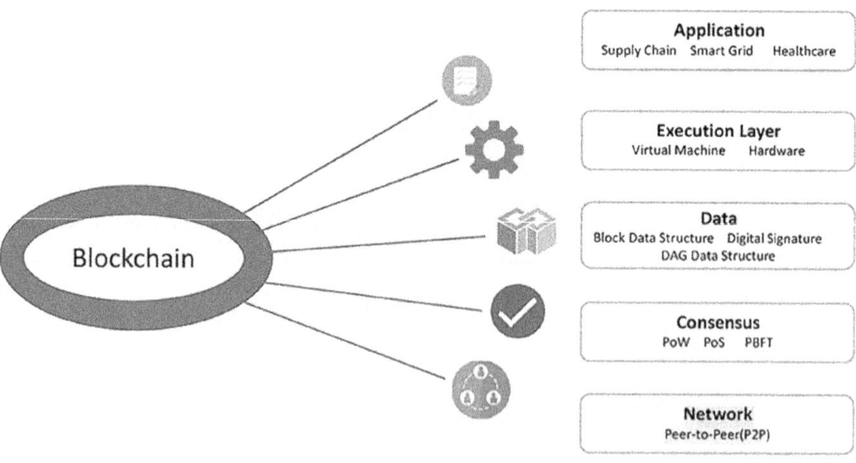

Figure 3.3 **A road map for the healthcare industry powered by blockchain.**

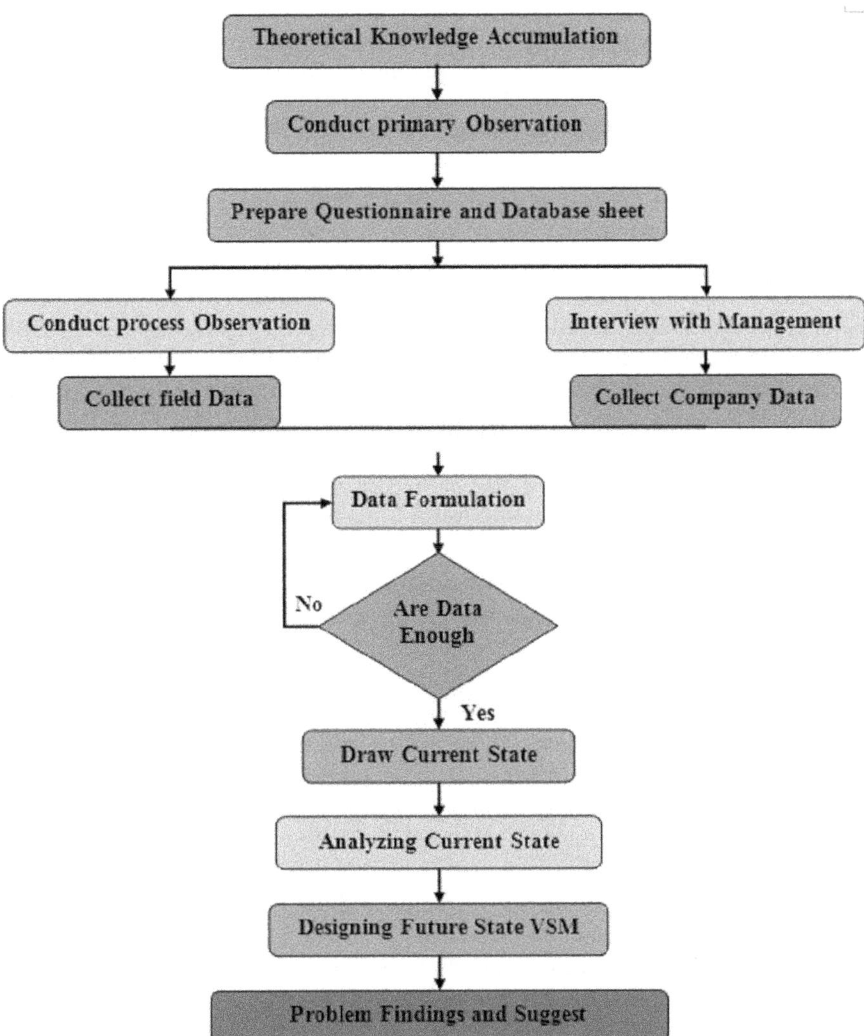

Figure 3.4 Research methodology flow diagram for research approach.

as well as a list of ways in which they affect medical care associations' daily operations. The results also indicate a deficiency of research and application-based activity in the field.

3.4 RESULTS AND DISCUSSION

Using blockchain innovation, validity and findings are introduced to the clinical preclinical phases. On blockchain, these papers can be stored in the digital thumbprint of an intelligent contract. A few advantages of applying blockchain technology in the healthcare sector are uniform authorization patterns for accessing electronic health information, identity verification and authentication for all parties involved, and network infrastructure security at every stage. The pharmaceutical supply chain is tracked and drug responsibility is managed via a blockchain. This technology aids in the analysis and confirmation of procedure results because it has the ability to save data about specific patients. Blockchain is used to improve safety, information presentation, and transparency, as well as for clinical research, patient monitoring, and medical record maintenance. It keeps hospital financial statements accurate while cutting down on the time and expense associated with data transformation. It addresses a few issues in the context of information-focused environments. Every block of patient health records will have a hash produced by blockchain innovation. Additionally, patients will be encouraged by the blockchain approach to share their personal information with third parties while keeping their identities private. Many informational indices are expected to do a clinical preliminary. The specialists focus on these informational indicators and regularly do experiments to offer analyses, assessments, and productivity ratios in a range of conditions. After the data is analyzed, more judgments are taken in view of the results. Nonetheless, a wide range of researchers possess the capability to modify the collected data and prove to change the outcome.

Many pharmaceutical businesses also wish to record the outcomes that will support their business operations. Thus, researchers are turning to blockchain technology to expedite and ensure fairness in clinical studies. It will simplify, standardize, and increase the security of clinical trial recording. Patient care might be enhanced and efficiency gains could be maximized by post-market analysis with the correct data. Blockchain technology, which provides improved privacy and security, resilience, open management, transparent auditing tracks, and data transparency, forms the foundation of these standards. This enables medical professionals to follow the most recent healthcare laws, including those pertaining to the safety of pharmaceutical supplies.

This new topic focuses mostly on the advantages of applying blockchain technology to the healthcare industry as well as the outstanding problems that prevent its widespread use. A thorough explanation is provided of the pros and pitfalls of blockchain use for healthcare businesses (Figure 3.5).

- **Motivations:** The result of modern society's efforts to meet needs in a variety of healthcare-related applications is the development of blockchain technology. Blockchain technology makes it possible to effectively increase patient quality without sacrificing goals related to system security. An analysis of the research is done to look into, categorize, and pinpoint the various benefits and justifications for applying blockchain technology in the healthcare industry. More discussion demonstrates these motivational categories.

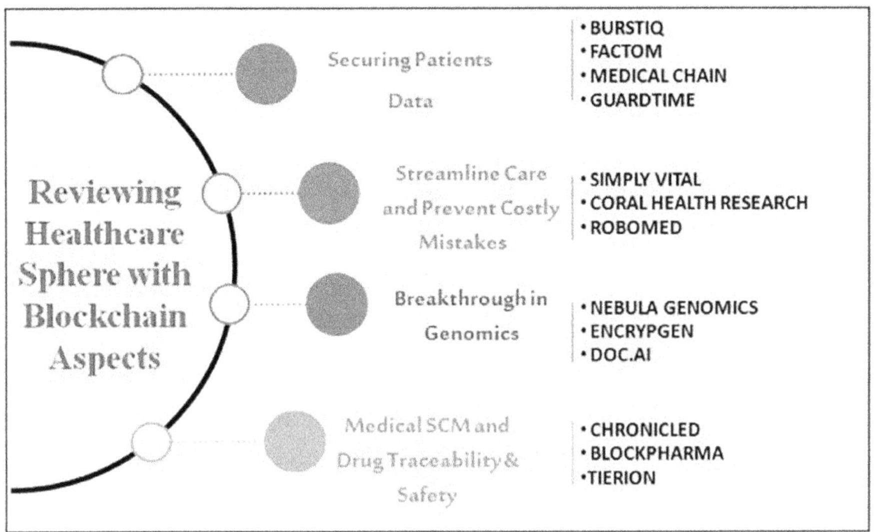

Figure 3.5 Blockchain in healthcare.

- **Decentralization:** Since medical data is distributed throughout the network rather than at a single point of security failure, the implementation of blockchain technology offers significant advantages for medical data. All parties engaged in the medical care industry must have consistent, secure, and immediate access to this data due to the decentralized accountability for data that this biological system takes into consideration. Additionally, this method allows medical data to be managed and transmitted under the control of an algorithm that generates a consensus mechanism based on input from dependable network users. A decentralized network has taken the place of the previous healthcare ecosystem, which comprised Remote Patient Mointoring (RPM), teledermatology, telesurgery, EHRs, EMRs, and Personal Health Record(PHR) systems. By resolving numerous problems, including those involving patient records, the interchangeability of medical data, and the security of healthcare facilities and services, this transition has greatly benefited the healthcare industry.
- **Transparency:** Thanks to smart contracts, the transaction- and multilateral interaction-based transparency feature of the blockchain is now more precise, reliable, and effective. This component addresses the opaqueness around patient verification in the healthcare industry by offering robust communication and information-sharing opportunities among various healthcare providers. The patient–specialist interchange is transferred through

a cunning agreement that permits the patient to speak with the specialist before making an arrangement. Beyond the blockchain, data transparency guarantees drug confidence, which enhances clinical information security and modifies its proofing.

- **Security and Privacy:** The volume of medical data is increasing, necessitating the need for novel processing and storage techniques. The use of blockchain technology demonstrates that safe data transit and storage in the medical industry is possible. Nodes connected to a network in blockchain technology are safeguarded by cryptography. By preventing manipulation, the SHA-256 hash function algorithm provides blockchains with data integrity security. A digital data checksum created by cryptographic hashes is a powerful feature that effectively protects the privacy of health data by prohibiting anyone from removing any information from the blockchain without permission.

- **Challenges:** Blockchain technology has several benefits, including the ability to assist the medical industry in resolving its numerous security, protection, and record-sharing issues. Nevertheless, before implementing blockchain technology directly in the medical services industry, there are certain disadvantages to take into account. Adaptability and capacity constraints, blockchain size, normalization, and ability associations are a few of these problems. This section examines and categorizes the difficulties that blockchain technology brings and shows the many task classifications.

  - **Scalability and Storage Capacity Issues:** Two primary issues arise when medical data is kept on the blockchain: flexibility and classification. The information on the blockchain is visible to every member of the organization due to its decentralized structure and simplicity, which renders it impractical for some applications in the medical services industry. Database storage may be significantly impacted by the data kept on the blockchain, which includes the patient's test results, MRIs, X-rays, and medical history.

  - **Blockchain Size Issues:** A sensor device connects patients to blockchain-based transactions, like RPM and EHRs, necessitating an increasing number of miners. The amount of data that Internet of Medical Things (IOMT) devices leak is too much for blockchain technology to handle.

  - **Universal Interoperability and Standardization Issues:** As blockchain is still in its infancy and is evolving quickly, there isn't a set standard for accessibility just yet. Owing to the requirement for universally assured normalization, the association would also have to devote additional time and resources to the application of blockchain technology in the healthcare industry. It would be advantageous for the standard permit if there was agreement on the types, sizes, and structures of data that might be kept on the

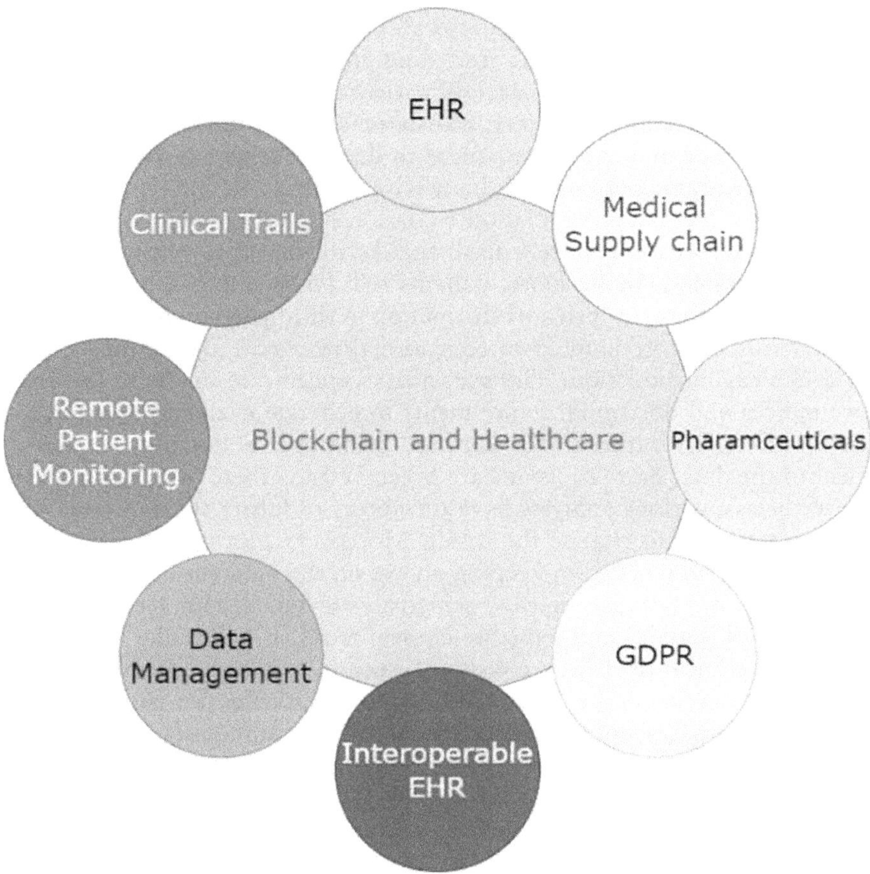

Figure 3.6 Applications of blockchain technology in healthcare.

blockchain. If blockchain technology was built on current standards that businesses could easily apply, adoption would increase.

- **Healthcare Organization Skill Issues**: The idea of a blockchain innovation action plan is not widely known. Hospitals and other healthcare organizations would need a considerable amount of time to completely convert their old RPM, EHR, PHR, and EMR infrastructure to blockchain technology (Figure 3.6).

Clinical preliminary testing gains validity and results via blockchain innovation. These records can be stored within the digital thumbprint of an intelligent contract on a blockchain. Consistent permission processes for accessing electronic health data, member character validation and verification, and extensive organization framework security are a few benefits of incorporating blockchain innovations in medical services. The network of pharmaceutical outlets is verified, and prescription liabilities are tracked using a blockchain. This technology can be utilized to preserve individual patient data, which aids in the study and validation of surgery outcomes. Blockchain is applied in clinical trials, patient monitoring, and health record retention. It also enhances safety, transparency, and information display. It reduces the time and expense involved in data transformation while maintaining the accuracy of hospital financial accounts.

It solves a few issues in relation to data centers. Every block of patient prosperity records will have a hash thanks to the development of blockchain technology. Furthermore, patients will be encouraged by the blockchain method to reveal private information to third parties while still being anonymous. A large number of educational resources are meant to serve as a clinical introduction. The specialists concentrate on these informative indices and do regular experiments to generate evaluations, analyses, and productivity ratios under various conditions. Following the assessment of the data, more decisions are taken keeping these findings in mind. Nevertheless, various analysts have the ability to falsify the data and verification gathered to change the result. Moreover, a lot of pharmaceutical companies are interested in keeping an eye on the outcomes that may help their company. Because of this, scientists use blockchain technology to guarantee objectivity and expedite clinical trials. It will make it easier to document clinical trials in a consistent, secure, and straightforward manner. Patient care could be improved by optimizing efficiency gains through post-market analysis of the collected data. These standards are based on improved privacy and security, open management, transparent audit trails, robustness, and openness of data that blockchain technology offers. As a result, healthcare providers are able to adhere to the most recent patient care criteria, which include safeguarding pharmaceutical supplies (Figure 3.7).

3.5 CONCLUSION

Blockchain's intrinsic encryption and decentralization enable innovative uses in the medical industry. It permits the production of counterfeit medications for use in conflict, encourages the adaptation of health data, improves interoperability among medical care organizations, and fortifies the security of patients' electronic clinical information. Blockchain technology has the power to completely transform a number of industries that provide medical services. The ability to facilitate complex arrangements made feasible by astute agreements in sectors like medical services is one of blockchain's most significant applications. Expenses will decrease because smart agreements cut off middlemen from the installment chain. The potential of blockchain in the healthcare industry is greatly impacted by the ecosystem's adoption of related cutting-edge technology. Clinical trials, health insurance, and system monitoring are all included. Hospitals are able to

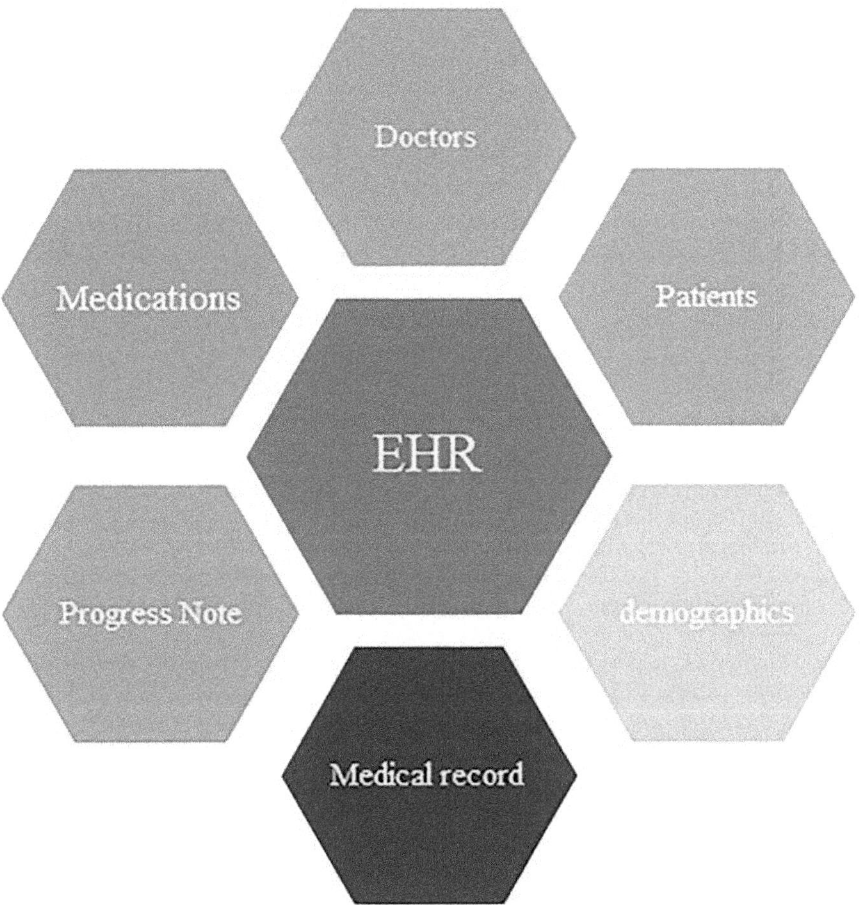

Figure 3.7 Electronic health record.

map out their services using a blockchain framework by utilizing device tracking during its whole life cycle. Blockchain technology can help executives' patient histories grow, which will speed up healthcare operations and improve information support, particularly in the protection intervention phase. Generally speaking, this innovation would greatly enhance and ultimately transform the way that medical care administrations are handled, put to use, and organized for both doctors and patients. Blockchain technology has the ability to completely change entire sectors. It could be able to improve security and make it harder to compromise the existing systems. The medical services sector is one where information is developing at a high rate. To improve healthcare, technologies like blockchain are required to store data securely, allow for analysis, and streamline the process of effectively recording information. The medical services sector has a great opportunity to embrace blockchain technology and spur innovation. The proposed endeavor entailed introducing blockchain technology to the medical industry.

This work is limited by the databases we searched. The study's chronology has also been affected by the rise in blockchain-related activity in the medical field. On the contrary, this study aims to evaluate the extensive blockchain research that has previously been conducted on the healthcare sectors to determine the gaps that remain between them. Blockchain technology is being studied by many academics in the healthcare sector.

REFERENCES

1. Haleem, A., Javaid, M., Singh, R. P., Suman, R., & Rab, S. (2021). Blockchain technology applications in healthcare: An overview. *International Journal of Intelligent Networks*, 2, 130–139.
2. Attaran, M. (2022). Blockchain technology in healthcare: Challenges and opportunities. *International Journal of Healthcare Management*, 15(1), 70–83.
3. Hussien, H. M., Yasin, S. M., Udzir, N. I., Ninggal, M. I. H., & Salman, S. (2021). Blockchain technology in the healthcare industry: Trends and opportunities. *Journal of Industrial Information Integration*, 22, 100217.
4. Mohammed, R., Alubady, R., & Sherbaz, A. (2021, March). Utilizing blockchain technology for IoT-based healthcare systems. *Journal of Physics: Conference Series*, 1818(1), 012111.
5. Johari, R., Kumar, V., Gupta, K., & Vidyarthi, D. P. (2022). BLOSOM: BLOckchain technology for security of medical records. *ICT Express*, 8(1), 56–60.
6. Attaran, M. (2022). Blockchain technology in healthcare: Challenges and opportunities. *International Journal of Healthcare Management*, 15(1), 70–83.
7. Hussien, H. M., Yasin, S. M., Udzir, N. I., Ninggal, M. I. H., & Salman, S. (2021). Blockchain technology in the healthcare industry: Trends and opportunities. *Journal of Industrial Information Integration*, 22, 100217.
8. Ahmad, R. W., Salah, K., Jayaraman, R., Yaqoob, I., Ellahham, S., & Omar, M. (2021). The role of blockchain technology in telehealth and telemedicine. *International Journal of Medical Informatics*, 148, 104399.
9. Prybutok, V. R., & Sauser, B. (2022). Theoretical and practical applications of blockchain in healthcare information management. *Information & Management*, 59(6), 103649.
10. Gimenez-Aguilar, M., De Fuentes, J. M., Gonzalez-Manzano, L., & Arroyo, D. (2021). Achieving cybersecurity in blockchain-based systems: A survey. *Future Generation Computer Systems*, 124, 91–118.
11. Odeh, A., Keshta, I., & Al-Haija, Q. A. (2022). Analysis of blockchain in the healthcare sector: Application and issues. *Symmetry*, 14(9), 1760.

12. Jadav, N. K., Rathod, T., Gupta, R., Tanwar, S., Kumar, N., & Alkhayyat, A. (2023). Blockchain and artificial intelligence-empowered smart agriculture framework for maximizing human life expectancy. *Computers & Electrical Engineering*, 105, 108486. ISSN 0045-7906. https://doi.org/10.1016/j.compeleceng.2022.108486.

13. Zulkifl, Z., Khan, F., Tahir, S., Afzal, M., Iqbal, W., Rehman, A., ... & Almuhaideb, A. M. (2022). FBASHI: Fuzzy and blockchain-based adaptive security for healthcare IoTs. *IEEE Access*, 10, 15644–15656.

14. Raju, K., Ramshankar, N., Shathik, J. A., & Lavanya, R. (2023). Blockchain assisted cloud security and privacy preservation using hybridized encryption and deep learning mechanism in IoT-healthcare application. *Journal of Grid Computing*, 21(3), 45. https://doi.org/10.1007/s10723-023-09678-7.

15. Haque, R., Sarwar, H., Kabir, S. R., Forhat, R., Sadeq, M. J., Akhtaruzzaman, M., & Haque, N. (2021). Blockchain-based information security of electronic medical records (EMR) in a healthcare communication system. In *Intelligent Computing and Innovation on Data Science: Proceedings of ICTIDS 2019* (pp. 641–650). Singapore: Springer Singapore.

## Chapter 4

Enhancing precision farming through IoT

Advanced sensor applications and data strategies in agriculture

Carmel Mary Belinda, E. Kannan, S. Sivagami, and Joy Winston James

4.1 INTRODUCTION

Precision farming, sometimes referred to as precision agriculture, is a cutting-edge method of farming that maximises agricultural practices by utilising technology [1]. This strategy seeks to minimise farming's negative environmental effects, increase agricultural yields, and cut down on resource waste. The Internet of Things (IoT) is one of the main technology elements of precision farming [2]. It is essential for gathering and organising data so that decisions can be made with knowledge. Although conventional agricultural practices have proven successful for centuries, a more effective and sustainable approach to agriculture is required due to the world's expanding population and shifting environmental conditions. To monitor, analyse, and optimise many aspects of agricultural production, precision farming uses technology. On the farm, real-time data collection and analysis from a variety of sources is made possible by IoT technology.

Precision farming overcomes difficulties such as limited water availability and erosion of soil by utilising IoT technologies. Through the utilisation of sensors and intelligent irrigation systems, it enhances water efficiency, preserves resources, and alleviates the consequences of water scarcity on crop yield. In addition, the implementation of precision farming methods, such as variable rate technology and soil monitoring, aids in the prevention of soil deterioration by accurately administering inputs and preserving soil vitality. The utilisation of remote sensing and artificial intelligence (AI) enables the prompt identification of problems, leading to timely solutions and improved sustainability. Precision farming is essential for reducing the impact of water scarcity and soil degradation and promoting sustainable agricultural practices in the long term.

Farmers are able to minimise their environmental effects, enhance production, lower input costs, and make well-informed decisions thanks to this data-driven strategy. The demand for creative agricultural solutions is greater than ever, as the world's population continues to rise and environmental

DOI: 10.1201/9781032618173-5

conditions undergo rapid shifts. Precision farming presents itself as a ray of hope, utilising technology to transform conventional farming methods. The IoT, a network of connected devices that makes data collecting, management, and decision-making easier, is a key enabler of precision farming. The investigation also includes the use of cutting-edge technologies like IoT soil moisture monitors, drones, weather stations, and remote sensing satellites [3]. When carefully considered, these technologies play a major role in large-scale farm management, crop monitoring, and irrigation optimisation. Precision farming increases productivity while fostering resource efficiency and environmental sustainability using these advancements.

A thorough grasp of real-time sensor data and remote sensing data from drones and satellites is provided by the methodical investigation of data collecting and transmission processes. The significance of software and algorithms in the precision farming ecosystem is highlighted by emphasising their transformative function in transforming raw data into usable insights. A variety of data transmission technologies are examined, including satellite communication, LoRaWAN, cellular networks, and Wi-Fi. Selecting the right protocols is essential for handling the massive and constant stream of data produced by precision agricultural systems [4]. The conversation also touches on the data security, device security, network security, and data privacy issues that are inherent in IoT systems for agriculture. Techniques like firmware updates, encryption, access controls, and compliance with data privacy laws are discussed in detail to protect sensitive agricultural data in an increasingly networked environment.

4.2 PRECISION FARMING WITH IoT SENSORS

Precision farming, which uses technology to maximise agricultural practices, has emerged as a revolutionary technique in the rapidly changing field of agriculture. IoT sensors are essential to the success of precision farming because they allow farmers to gather vital data for well-informed decision-making. This chapter explores the critical role that IoT sensors play in precision farming, emphasising the use of weather stations, drones, soil moisture sensors, and remote sensing satellites.

The several sensors built into drones for precision farming are displayed in Figure 4.1. LiDAR sensors, GPS receivers, thermal cameras, and multispectral cameras are some of the tools that help with resource management, yield prediction, field mapping, pest and disease detection, and crop health monitoring [5].

4.2.1 Sensors of soil moisture

4.2.1.1 Soil Moisture's Significance in Agriculture

One must first comprehend the crucial role soil moisture plays in crop growth to appreciate the significance of soil moisture sensors in precision

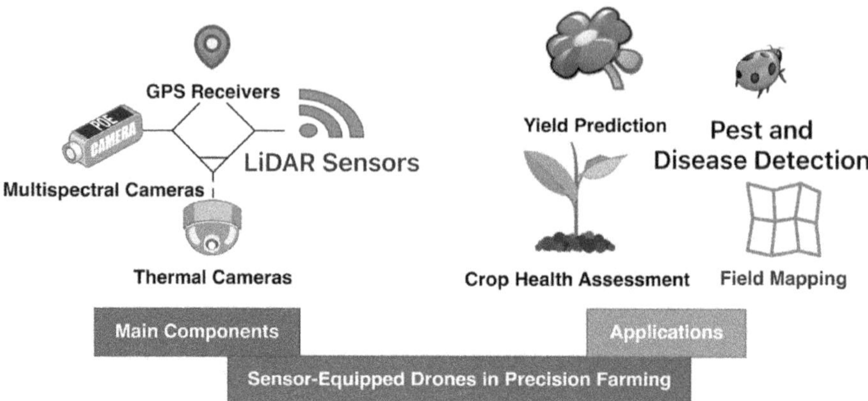

Figure 4.1 Sensor-equipped drones in precision farming.

farming. The amount of moisture in the soil has a direct effect on how well a plant absorbs water and nutrients. While too much rainfall can result in overwatering, too little moisture can create drought stress, both of which are harmful to the health and productivity of crops.

4.2.1.2 Soil Moisture Sensors Powered by IoT

Real-time information on soil moisture levels at different depths in the soil profile is intended to be provided via IoT-enabled soil moisture sensors [6]. These sensors use a range of technologies to provide precise readings; capacitance and resistance measurements are two popular techniques.

Capacitance soil moisture sensors use variations in electrical capacitance between two in-soil electrodes to determine the moisture content of the soil. Soil moisture causes variations in its dielectric constant, which in turn causes changes in capacitance. A moisture reading is subsequently obtained from this change. Because of their reputation for accuracy, capacitance sensors are frequently employed in applications related to precision farming.

By measuring the electrical resistance between two probes put into the soil, resistance-based sensors calculate the moisture content of the soil. Lower resistance is the result of the soil's increased electrical conductivity with increasing wetness. Higher moisture levels are correlated with this decrease in resistance. Resistance sensors are renowned for being easy to use and reasonably priced.

The level of moisture in the soil is crucial for the growth of plants as it has a direct impact on the absorption of nutrients, the process of photosynthesis, and the overall well-being of the plant. Optimal soil moisture levels guarantee that plants receive enough amount of water for their metabolic processes, while also allowing for optimum root aeration. Inadequate moisture results in the withering of plants, hindered growth, and decreased crop production, whereas excessive moisture can lead to root decay and the leaching of nutrients. Precision farming methods, such as utilising IoT technologies to monitor soil moisture, allow farmers to effectively manage water levels, which in turn promotes robust crop development, maximises yields, and supports sustainable agriculture by saving water resources and minimising environmental harm.

4.2.1.3 Utilisation in Accurate Agriculture

Several advantages are available to farmers who practise precision farming with IoT-enabled soil moisture sensors:

- **Optimised Irrigation:** Farmers can adjust their irrigation techniques to the unique requirements of their crops by regularly monitoring the levels of soil moisture. With this focused strategy, irrigation expenses are decreased, water waste is minimised, and crops are given the ideal quantity of moisture for growth.
- **Prevention of Drought Stress:** Soil moisture sensors alert farmers to soil moisture shortages early, enabling them to take immediate corrective action. This proactive strategy aids in preventing drought stress, which can lower yields and limit crop growth.
- **Data-Driven Decision-Making:** By integrating sensor data with farm management software, farmers may use data to inform decisions about resource allocation and irrigation scheduling.

4.2.2 Meteorological stations

4.2.2.1 How Weather Affects Agriculture

A continuous and dynamic aspect of agriculture is the weather. Important factors that affect crop growth, the frequency of pests and diseases, and overall farm management decisions are temperature, humidity, wind speed, rainfall, and sun radiation.

4.2.2.2 IoT Weather Stations

Precision farming requires the use of IoT weather stations. Farmers can make educated decisions with the help of a wide range of sensors these stations are outfitted with, which collect vital meteorological data [7]. Typical sensors seen in IoT weather stations are:

- **Temperature Sensors:** These devices take air temperature readings, which are essential for comprehending the growth and development of plants.
- **Humidity Sensors:** Humidity sensors measure the amount of moisture in the air, offering information on the state of the atmosphere and possible dangers of sickness.

- **Wind Speed and Direction Sensors:** The management of spray applications and the estimation of the danger of windborne disease transmission depend on wind data.
- **Rainfall Sensors:** Farmers can monitor precipitation levels, determine the availability of water, and adjust irrigation with the use of rainfall data.
- **Solar Radiation Sensors:** These sensors track the quantity of sunlight that reaches the surface of the planet, which helps with estimates of energy generation and crop growth forecasts.

For example, consider a scenario where a farmer has a weather station equipped with IoT technology put on their land. This station continuously collects data on temperature, humidity, and precipitation. Through the analysis of this data, the farmer can ascertain the optimal periods for sowing seeds and applying irrigation. For example, if the weather station forecasts a period of high temperatures and low precipitation, the farmer can choose to postpone planting until conditions become more advantageous, to guarantee that the seeds have sufficient moisture for germination. Likewise, in the event that the station predicts a significant amount of rain, the farmer may choose to delay irrigation to prevent the soil from being waterlogged. IoT weather stations assist farmers in making well-informed decisions to enhance crop development and maximise production.

4.2.2.3 Utilising Precision Agriculture

Precision farming techniques are significantly impacted by the information gathered by IoT weather stations:

- **Planting and Harvesting Decisions:** To maximise crop yield and quality, farmers can utilise temperature and sun radiation data to identify the best times to plant and harvest.
- **Disease and Pest Management:** By anticipating disease outbreaks and insect infestations, knowledge of temperature and humidity conditions enables prompt intervention.
- **Resource Allocation:** Farmers can make decisions about irrigation, fertilisation, and pest management with the use of rainfall data.
- **Adaptation to Climatic Change:** Weather stations offer insightful information on regional climatic patterns, allowing farmers to modify their operations in response to shifting circumstances.

4.2.3 Unmanned aerial vehicles

4.2.3.1 Agriculture's Drone Revolution

Unmanned aerial vehicles, or UAVs, sometimes known as drones, have brought agriculture into a new era [8]. Their capacity to offer a thorough, high-resolution aerial perspective of the farm has revolutionised disease diagnosis, yield prediction, and crop monitoring.

4.2.3.2 Drones with Sensors

Drones with IoT sensors built in are effective instruments for precision farming. Among these sensors are the following:

- **Multispectral Cameras:** By taking pictures at different wavelengths, these cameras make it possible to evaluate the health of crops, nutrient shortages, and the presence of diseases.
- **Thermal Cameras:** Using thermal images to detect temperature differences throughout the field, problems like water stress or drainage issues can be found.
- **GPS Receivers:** Accurate data collection and analysis are made possible by GPS technology, which allows for precise mapping of the field.
- **LiDAR Sensors:** These sensors, which stand for light detection and ranging, produce intricate three-dimensional maps of the landscape that help with drainage planning and topographic analysis.

4.2.3.3 Utilising Precision Agriculture

Precision farming applications for drones with IoT sensors are numerous and varied.

- **Crop Health Assessment:** Using multispectral photography, farmers may target interventions like fertilisation and pest management by identifying changes in crop health.
- **Insect and Disease Identification:** Using thermal imaging, regions of the field with unusual temperature patterns can be found, which may be a sign of insect or disease infestations.
- **Yield Prediction:** Drones can help farmers plan for harvest by gathering data during the growing season and contributing to precise yield projections.
- **Field Mapping:** Accurate field maps are made possible by GPS and LiDAR data, which also helps to optimise drainage and planting schemes.
- **Resource Management:** By offering a comprehensive perspective of the farm, drones facilitate effective resource allocation and decision-making.

4.2.4 Satellites for remote sensing

Precision Farming with Satellite Technology: The earth's orbiting remote sensing satellites provide a macroscopic view of agricultural fields [9]. They are essential equipment for precision farming since they offer vital information on vegetation indices, soil moisture, and crop health.

4.2.4.1 Satellite-Captured Data from Remote Sensing

Remote sensing satellites use a variety of sensors and imaging technologies to collect a vast amount of data, such as:

- **Multispectral Imaging:** By collecting information at various wavelengths, these sensors enable the evaluation of crop stress and health.
- **Synthetic Aperture Radar (SAR):** SAR sensors can see through clouds to get data on topography and soil moisture.
- **Vegetation Indices:** Information derived from these indices can reveal pest or disease infestations, plant health, and growth stages.

Satellite imaging facilitates the efficient management of huge agricultural estates with its wide-ranging coverage and remote monitoring capabilities. It provides immediate analysis of crop health and environmental factors, allowing for early identification of problems such as pest infestations and water scarcity. This data enables the use of precise management techniques, which in turn optimise the utilisation of resources and maximise crop production. Satellite imaging, when combined with other data sources, offers a full perspective on the health of farms. This helps in implementing cost-effective and sustainable management plans for large-scale agriculture.

Use Case: Accurate Watering Using Soil Moisture Indicators
- **Input:**
  - **Data from Soil Moisture Sensors:** Current information gathered from sensors positioned at different soil profile depths.
  - **Crop Information:** Crop type, stage of growth, and particular water needs.
  - **Meteorological Data:** Present and anticipated meteorological conditions, such as temperature and precipitation.
- **Method:**
  - **Data Gathering:** Soil moisture sensors continuously track the soil's moisture content at various depths. A central system receives the data transmission.
  - **Crop and Weather Integration:** The data on soil moisture is combined with details on the particular crop being grown and the weather at the moment by the central system.
  - **Decision-Making Algorithm:** An algorithm processes the combined data to calculate the ideal soil moisture content required for a specific crop under the given weather conditions. A graphical representation of soil moisture levels at different depths is provided in Figure 4.2.

Sample Code:

```python
import random
import matplotlib.pyplot as plt

def collect_soil_moisture_data():
    # Simulate real-time soil moisture data collection
    depths = ['0-10cm', '10-20cm', '20-30cm']
```

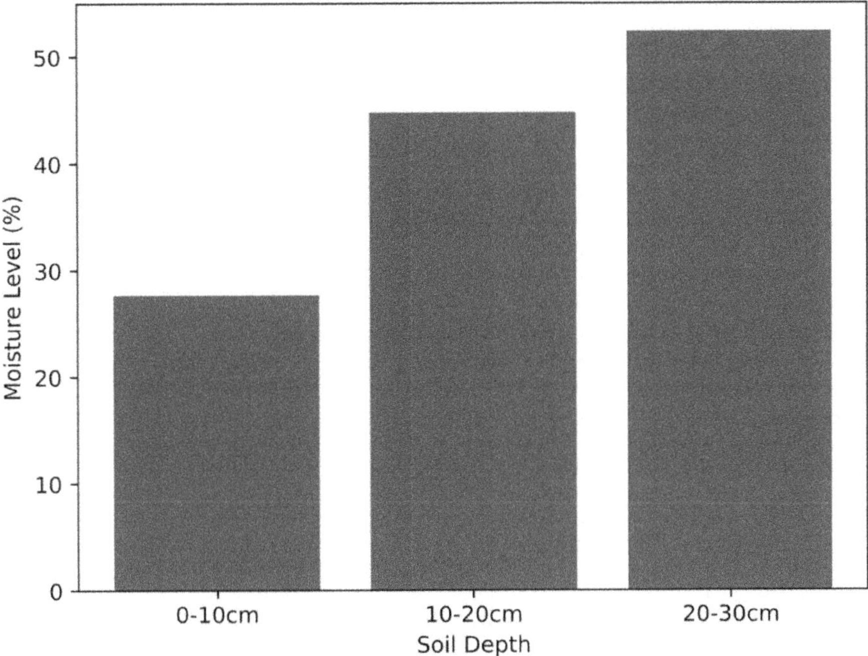

Figure 4.2 Soil moisture levels at different depths.

```
    moisture_levels = {depth: random.uniform(20, 60) for
depth in depths}
    return moisture_levels
def plot_soil_moisture_data(moisture_levels):
    # Plot soil moisture levels
    depths = list(moisture_levels.keys())
    levels = list(moisture_levels.values())
    plt.bar(depths, levels, color='blue')
    plt.title('Soil Moisture Levels at Different Depths')
    plt.xlabel('Soil Depth')
    plt.ylabel('Moisture Level (%)')
    plt.show()

def integrate_crop_and_weather_data(crop_info,
weather_data):
    # Simulate integration of crop and weather data
    integrated_data = {
        'crop_type': crop_info['crop_type'],
        'growth_stage': crop_info['growth_stage'],
        'water_requirements':
crop_info['water_requirements'],
        'temperature': weather_data['temperature'],
        'precipitation': weather_data['precipitation']
    }
```

```python
    return integrated_data
def decision_making_algorithm(integrated_data):
    # Simulate a simple decision-making algorithm
    optimal_soil_moisture = integrated_data['water_
requirements'] - 10
    return optimal_soil_moisture

def generate_irrigation_schedule(optimal_soil_moisture):
    # Simulate irrigation schedule generation
    schedule = f"Irrigate when soil moisture is below
{optimal_soil_moisture}%"
    return schedule

def provide_watering_recommendations(schedule):
    # Simulate providing watering recommendations
    recommendations = f"Follow the irrigation schedule:
{schedule}"
    return recommendations

def main():
    # Sample input data
    soil_moisture_data = collect_soil_moisture_data()
    crop_information = {'crop_type': 'Tomatoes', 'growth_
stage': 'Vegetative', 'water_requirements': 50}
    weather_data = {'temperature': 28, 'precipitation': 'Sunny'}

    # Plot soil moisture data
    plot_soil_moisture_data(soil_moisture_data)

    # Process data
    integrated_data = integrate_crop_and_weather_data(crop_
information, weather_data)
    optimal_soil_moisture =
decision_making_algorithm(integrated_data)
    irrigation_schedule =
generate_irrigation_schedule(optimal_soil_moisture)
    watering_recommendations =
provide_watering_recommendations(irrigation_schedule)

    # Sample output
    print("Sample Output:")
    print(f"Optimized Irrigation Schedule:
{irrigation_schedule}")
    print(f"Watering Recommendations:
{watering_recommendations}")

if __name__ == "__main__":
    main()
```

Sample Output:

Optimised Irrigation Schedule: Irrigate when soil moisture is below 40%.

- Watering Recommendations: Follow the irrigation.
- Schedule: Irrigate when soil moisture is below 40%.

Result:

Optimised Irrigation Plan: Based on crop requirements, weather forecasts, and real-time soil moisture data, the system creates an optimised irrigation plan.

Watering Recommendations: To guarantee that irrigation is customised to the exact requirements of the plants, farmers receive practical advice on when and how much water to apply to their crops.

Resource Savings: Precision irrigation lowers irrigation costs, conserves water, and encourages effective resource use.

Drought Stress Prevention: Proactive steps can be taken to prevent crop damage and yield loss by receiving early indicators of possible drought stress.

Data Logging and Analysis: Farmers can monitor trends over time and improve their irrigation plans for upcoming seasons using the system, which logs all data for historical analysis.

This use case illustrates how soil moisture sensors can support precision irrigation techniques when combined with crop and meteorological data. A customised irrigation plan that maximises water use, fosters crop health, and reduces farming's negative environmental effects is the result.

4.3 PRECISION FARMING: DATA GATHERING AND TRANSFER

Data is essential to precision farming because it facilitates well-informed decision-making that results in improved productivity and optimised agricultural practices [10]. This chapter explores the complex network of cloud-based data storage techniques, transmission protocols, data integration, processing, and gathering strategies used in precision farming. We examine how these components work together to provide farmers with the information they need to effectively run their businesses.

4.3.1 Techniques for gathering data

Precision farming is centred on data collecting, which serves as the cornerstone for all ensuing analysis and decision-making. Real-time data and distant sensing data are the two main categories into which we group data collecting in this section [11]. An example of how real-time data on soil conditions, meteorology, and crop monitoring is provided via soil moisture sensors, weather stations, and in-field cameras is shown in Figure 4.3.

4.3.1.1 Data in Real Time

Real-time data gathering entails using a variety of sensors and equipment to continuously monitor and record agricultural conditions. This kind of

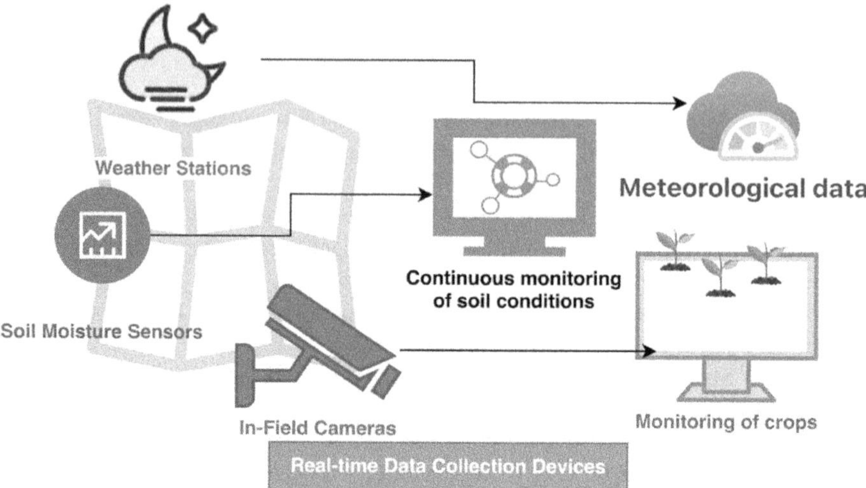

Figure 4.3 Real-time data collection devices.

data collecting allows for quick responses to changing situations by providing instant insights into the farm's condition. Several primary sources of up-to-date information consist of:

- **Soil Moisture Sensors:** As covered in Chapter 2, soil moisture sensors help farmers avoid overwatering or drought stress by continuously measuring soil moisture levels at various depths.
- **Weather Stations:** IoT weather stations offer current meteorological information, such as temperature, humidity, wind direction and speed, precipitation, and sun radiation. Making weather-informed decisions on the farm requires this data.
- **In-Field Cameras:** Real-time photos and videos are captured by strategically positioned cameras across the field. Crop growth, insect and disease infestations, and equipment performance can all be tracked using these cameras.

4.3.1.2 Data from Remote Sensing

Utilising tools like satellites and drones to gather data from a distance is known as remote sensing data collecting [12]. These resources offer a more comprehensive view of the farm as a whole as well as recurring snapshots of crop health and environmental circumstances. Primary sources of data from remote sensing include:

- **Drones:** Outfitted with an array of sensors, these aerial vehicles obtain detailed photographs and information from above the farm. They are used for yield prediction, pest detection, and crop monitoring. Drones can swiftly cover wide regions and reach difficult-to-reach areas of the farm.

- **Satellite Imagery:** Important information on crop health, soil moisture, and vegetation indices is provided by remote sensing satellites circling the earth. When it comes to pest control, fertilisation, and irrigation, the information they gather can be very helpful in making prompt decisions.

4.3.2 Processing and integration of data

Getting started with raw data collection is just the first step towards data-driven agriculture. The gathered data needs to be processed, analysed, and integrated to yield valuable insights. This section delves into the critical functions of software tools, algorithms, and integration platforms in converting data into knowledge that can be put to use.

- **Platforms for Integration:** By acting as the link between disparate data sources, integration platforms make sure that information moves smoothly to a central location where it can be handled and examined. These platforms make it possible to combine data from various sensors, drones, and satellite sources to get an all-encompassing picture of the farm's activities [13].
- **Software for Farm Management:** Software for farm management is essential for turning data into insights that can be put to use [14]. These software programmes provide instruments for data analysis, visualisation, and judgement. Dashboards that offer predictive analytics, historical data patterns, and real-time updates on agricultural conditions are available to farmers.
- **Machine Learning and Algorithms:** Patterns, trends, and anomalies are extracted from the data using algorithms and machine learning models [15]. These instruments can detect disease outbreaks, forecast crop yields, and allocate resources as efficiently as possible. Precision farming requires more and more machine learning as the amount and complexity of data increase.

Satellite imaging facilitates the efficient management of huge agricultural estates with its wide-ranging coverage and remote monitoring capabilities. It provides immediate analysis of crop health and environmental factors, allowing for early identification of problems such as pest infestations and water scarcity. This data enables the use of precise management techniques, which in turn optimise the utilisation of resources and maximise crop production. Satellite imaging, when combined with other data sources, offers a full perspective on the health of farms. This helps in implementing cost-effective and sustainable management plans for large-scale agriculture [16–19].

4.3.3 Protocols for data transmission

When it comes to precision farming, where prompt decisions can have a major impact on crop health and productivity, efficient and dependable data transmission is essential [20]. Many data transmission methods utilised by agricultural IoT sensors and equipment are examined in this section.

- **Wireless:** On farms, Wi-Fi is frequently used for data transmission, particularly in situations where sensors are close to a central hub or router [18]. It is appropriate for real-time data collecting and processing and offers fast data transport.
- **Networks of Cells:** Cellular networks provide a dependable method of data transfer in remote locations and on big farms with widely dispersed sensor deployments. Even in places without Wi-Fi service, data can be delivered to a central repository thanks to cellular connectivity.
- **Low-Power Wide-Area Network, or LoRaWAN:** For IoT devices that need low-power and long-range connectivity, LoRaWAN technology is perfect [16]. It provides extended coverage without rapidly depleting sensor batteries, making it ideal for farms with widely dispersed sensors.
- **Satellite-Based Linkage:** For data transfer in areas with poor terrestrial connectivity, satellite communication offers a backup plan. Because this approach guarantees data transmission regardless of the farm's location, it is especially beneficial for farms that are far away from one another.

4.3.4 Data storage via the cloud

Robust data storage systems are crucial in precision farming due to the massive volume of data generated. Scalability, accessibility, and security are provided by cloud-based storage for the enormous amounts of data gathered on the farm [17].

- **Flexibility:** Farmers can preserve historical data and do long-term trend analysis thanks to cloud-based storage options that can grow with their information.
- **Availability:** Remote management and decision-making are made easier for farmers and stakeholders because they may view their data from any location with an internet connection. This accessibility is particularly beneficial for companies that are spread out geographically and in size.
- **Protection of Data:** Cloud providers make significant investments in data security protocols to protect private agricultural data from hackers and unauthorised access. Cloud-based storage systems come equipped with encryption, access controls, and data redundancy as basic features.
- **Restoring and Backup Data:** Strong backup and recovery features are often provided by cloud platforms, guaranteeing that data is safeguarded even in the event of hardware malfunctions or other unanticipated events.

A whole ecosystem of data collection, integration, processing, transmission, and storage is necessary for precision farming [19]. These interrelated components give farmers the knowledge they need to allocate resources optimally, make wise decisions, and improve the productivity and sustainability of agricultural operations. Precision farming has the potential to transform agriculture and ensure that food production can meet global demand while minimising environmental effects as technology progresses.

Use Case: Precision Farming Optimisation through Data-Driven Decision-Making

This use case fully utilises the potential of precision farming by taking a data-driven approach to decision-making to optimise crop yield and resource management. First, a variety of data sources are gathered, such as records of pests and diseases, crop rotation patterns, soil health metrics, temperature, humidity, and rainfall, as well as historical crop production data. The collection is further enhanced with remote sensing data, which is obtained from drone-captured photos for crop health evaluation and satellite photography for vegetation indices. After that, the data is carefully processed to find and fix any missing or inconsistent points, and different datasets are combined into a coherent format for analysis.

Feature extraction is the following stage, when important indications are revealed, such as vegetation indices from satellite data and relationships between past weather patterns and agricultural productivity. Customised machine learning methods are utilised, such as regression models for yield forecasting, classification models for identifying pest and disease risks, and predictive models for ideal crop rotation. To make precise predictions, these models are trained on past data to identify patterns and linkages.

A key component of this use case is real-time decision support, where continuous data from weather stations, in-field cameras, and soil moisture sensors is analysed in real time for predictive analysis. The goal of the output is to enable farmers to make wise decisions. This includes suggestions for the most effective use of resources, such as targeted irrigation based on soil moisture content, dynamic fertilisation schedules customised for soil health, and early detection of pests and diseases with suggested treatments. Moreover, a dashboard visualisation offers farmers an easy-to-use interface together with visuals for risk variables, recommended resource allocation, and expected crop yields.

Finally, this use case highlights how powerful machine learning models combined with real-time and historical data integration may provide farmers with actionable insights, demonstrating the revolutionary potential of precision farming. The process's inherent continuous learning loop guarantees flexibility and advancement throughout time, eventually assisting in the creation of effective and sustainable farming methods.

Steps to demonstrate the use case:

Data Collection: Soil moisture sensors utilising IoT technology are strategically placed across the farm to continuously monitor and provide real-time updates on soil moisture levels. The sensors gather data on soil moisture at various sites within the fields on a continuous basis.

Integration with Weather Data: The soil moisture data acquired by the sensors is combined with weather forecasts obtained from meteorological stations or online sources to create an integrated system. This integration offers valuable information regarding future weather patterns, encompassing forecasts for precipitation and fluctuations in temperature.

Analysis and Processing: The farm management software utilises integrated data to analyse historical patterns and current circumstances. This analysis helps identify the specific water needs of different crops at different stages of growth.

Decision-Making: The software utilises studied data to develop watering regimens that are optimised for each crop's individual requirements and the current environmental circumstances. These schedules are created by taking into account factors such as soil type, crop type, weather forecasts, and past irrigation data.

Implementation: The watering schedules that are created are sent to irrigation systems that have IoT controllers. These controllers regulate the timing and length of irrigation according to predetermined schedules, guaranteeing that crops receive the appropriate amount of water at the optimal moment.

Monitoring and Feedback: The IoT soil moisture sensors continuously monitor soil moisture levels during the whole growth season. Alerts are triggered whenever there are any variations from the planned levels, which prompts farmers to examine and make adjustments to irrigation schedules if needed.

Assessment and optimisation: Farmers review the efficacy of watering regimens at the conclusion of the season by evaluating crop health, yield, and resource utilisation. Identifying any areas that need improvement and making adjustments optimises future watering plans (Figure 4.4).

Sample Code:

```python
# Import necessary libraries
import pandas as pd
from sklearn.model_selection import train_test_split
from sklearn.linear_model import LinearRegression
from sklearn.metrics import mean_squared_error
import matplotlib.pyplot as plt

# Mock dataset creation
data = {
    'Historical_Yield': [10, 12, 8, 15, 14],
    'Soil_Health': [60, 55, 70, 65, 75],
    'Temperature': [28, 30, 25, 32, 29],
    'Humidity': [60, 55, 65, 70, 75],
```

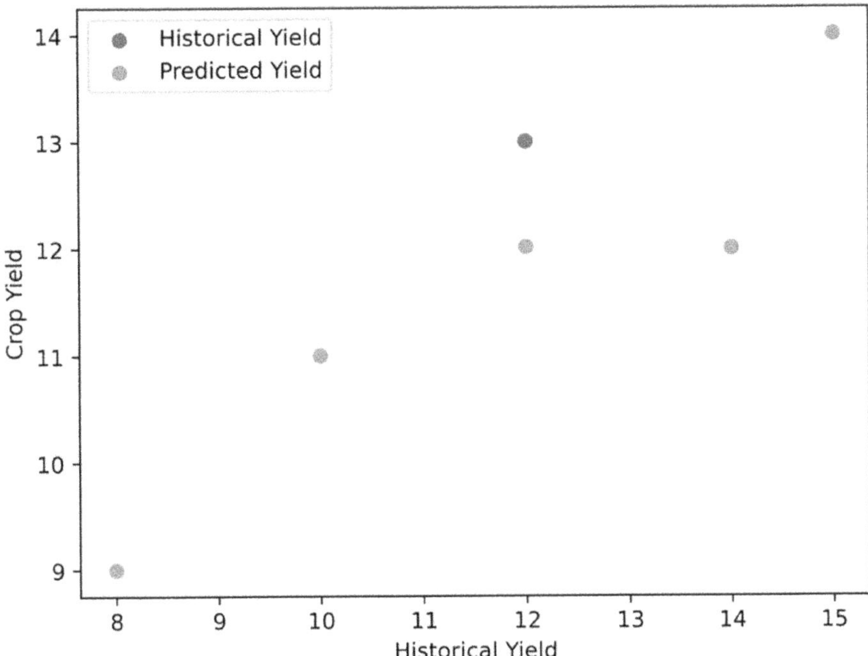

Figure 4.4 Historical yield versus predicted yield.

```python
    'Rainfall': [20, 15, 25, 10, 30],
    'Pest_Disease_Risk': [0.1, 0.2, 0.15, 0.25, 0.18],
    'Crop_Rotation_Factor': [0.8, 0.9, 0.7, 1.0, 0.85],
    'Crop_Yield': [11, 13, 9, 14, 12]
}

df = pd.DataFrame(data)

# Data Preprocessing
X = df.drop('Crop_Yield', axis=1)
y = df['Crop_Yield']

# Train-Test Split
X_train, X_test, y_train, y_test = train_test_split(X, y,
test_size=0.2, random_state=42)

# Machine Learning Model (Linear Regression in this case)
model = LinearRegression()
model.fit(X_train, y_train)

# Prediction and Optimization (Using a simplified real-time
input)
real_time_input = {
    'Historical_Yield': [10],  # Replace with the actual
historical yield
```

```
    'Soil_Health': [68],
    'Temperature': [27],
    'Humidity': [62],
    'Rainfall': [22],
    'Pest_Disease_Risk': [0.12],
    'Crop_Rotation_Factor': [0.88]
}

predicted_yield = model.predict(pd.DataFrame(real_time_
input, columns=X_train.columns))
# Output Visualization
print("Predicted Crop Yield:", predicted_yield[0])

# Feedback Loop (Continuous Learning can be implemented here
by updating the model with new data)

# Visualization (for simplicity, plotting the historical vs.
predicted yield)
plt.scatter(df['Historical_Yield'], df['Crop_Yield'],
label='Historical Yield')
plt.scatter(df['Historical_Yield'], model.predict(X),
label='Predicted Yield')
plt.xlabel('Historical Yield')
plt.ylabel('Crop Yield')
plt.legend()
plt.savefig('fgure 06.png',dpi=1000)
plt.show()

Predicted Crop Yield: 9.820860001787551
```

4.4 CONCLUSION

Precision farming is the cutting edge of contemporary agriculture and provides a revolutionary solution to the problems brought on by an expanding world population and changing environmental circumstances. An era of data-driven decision-making has arrived in agriculture because of the integration of IoT technologies, such as sensors, drones, and satellites. This allows farmers to maximise resource utilisation, improve production, and reduce environmental impact. Precision farming has several uses, including crop monitoring, precision irrigation, automated equipment, and predictive analytics. These examples highlight how versatile this technology is in transforming conventional agricultural methods. Real-time and remote sensing data collection, analysis, and interpretation enable farmers to make well-informed decisions that improve crop health, water use, and overall operational efficiency. Nonetheless, there are obstacles in the way of precision farming's broad acceptance. Stakeholders must work together to address issues including data security, implementation costs, technology integration, and environmental sustainability. Governments, business leaders, and academic institutions must work together to guarantee that farmers have access to the resources, funding, and legal protections they need to overcome these obstacles. With much promise, precision farming has a bright future. Advancements in machine learning, artificial intelligence, and sustainable agriculture are going to enhance and broaden the scope of precision farming systems. A more accessible, effective, and sustainable future for global agriculture will be shaped by the continued cooperation of researchers, farmers, and industry leaders as this sector develops.

IoT technologies such as sensors and drones play pivotal roles in achieving precision farming objectives. Sensors provide real-time data on soil moisture, nutrient levels, and weather conditions, enabling optimised resource utilisation and precise input applications. Drones offer aerial imagery for assessing crop health, detecting issues early, and implementing targeted interventions. Together, these technologies enhance soil management practices, minimise waste, and promote sustainable agriculture. By harnessing the power of IoT, precision farming maximises productivity while minimising environmental impact, ensuring a more resilient and efficient agricultural system for the future.

ACKNOWLEDGEMENTS

First and foremost, I would like to express my gratitude to co-authors whose expertise and guidance were invaluable in navigating the intricate landscape of IoT technologies in agriculture. I am also grateful for the editorial assistance, helping to refine and polish the manuscript for clarity and coherence. This chapter represents a collective effort. Thank you to everyone who played a role, no matter how small, in bringing this chapter to fruition. I declare that this work was not supported by the agency.

REFERENCES

1. Shaikh, T.A., Rasool, T. and Lone, F.R., 2022. Towards leveraging the role of machine learning and artificial intelligence in precision agriculture and smart farming. *Computers and Electronics in Agriculture*, *198*, p. 107119.
2. Khanna, A. and Kaur, S., 2019. Evolution of Internet of Things (IoT) and its significant impact in the field of precision agriculture. *Computers and Electronics in Agriculture*, *157*, pp. 218–231.
3. Maddikunta, P.K.R., Hakak, S., Alazab, M., Bhattacharya, S., Gadekallu, T.R., Khan, W.Z., & Pham, Q. V. 2021. Unmanned aerial vehicles in smart agriculture: Applications, requirements, and challenges. *IEEE Sensors Journal*, *21*(16), pp. 17608–17619.
4. Vuran, M.C., Salam, A., Wong, R. and Irmak, S., 2018. Internet of underground things in precision agriculture: Architecture and technology aspects. *Ad Hoc Networks*, *81*, pp. 160–173.
5. Fahey, T., Pham, H., Gardi, A., Sabatini, R., Stefanelli, D., Goodwin, I. and Lamb, D.W., 2020. Active and passive electro-optical sensors for health assessment in food crops. *Sensors*, *21*(1), p. 171.

6. Santos, G., Vicente, G., Salvado, T., Gonçalves, C., Caetano, F. and Silveira, C., 2023. Sustainable and autonomous soil irrigation: Agro smart solution. In Luísa Cagica Carvalho, Clara Silveira, Leonilde Reis, Nelson Russo, *Internet of Behaviors Implementation in Organizational Contexts* (pp. 359–379). Pennsylvania, USA: IGI Global.

7. Javaid, M., Haleem, A., Singh, R.P. and Suman, R., 2022. Enhancing smart farming through the applications of Agriculture 4.0 technologies. *International Journal of Intelligent Networks*, 3, pp. 150–164.

8. Boursianis, A.D., Papadopoulou, M.S., Diamantoulakis, P., Liopa-Tsakalidi, A., Barouchas, P., Salahas, G., Karagiannidis, G., Wan, S. and Goudos, S.K., 2022. Internet of things (IoT) and agricultural unmanned aerial vehicles (UAVs) in smart farming: A comprehensive review. *Internet of Things*, 18, p. 100187.

9. Ray, A.S., 2016. Remote sensing in agriculture. *International Journal of Environment, Agriculture and Biotechnology*, 1(3), p. 238540.

10. Malhotra, K. and Firdaus, M., 2022. Application of Artificial Intelligence in IoT Security for Crop Yield Prediction. *ResearchBerg Review of Science and Technology*, 2(1), pp. 136–157.

11. Cheng, T. and Teizer, J., 2013. Real-time resource location data collection and visualization technology for construction safety and activity monitoring applications. *Automation in construction*, 34, pp. 3–15.

12. Bansod, B., Singh, R., Thakur, R. and Singhal, G., 2017. A comparison between satellite based and drone based remote sensing technology to achieve sustainable development: A review. *Journal of Agriculture and Environment for International Development (JAEID)*, 111(2), pp. 383–407.

13. Garrett, B. and Anderson, K., 2018. Drone methodologies: Taking flight in human and physical geography. *Transactions of the Institute of British Geographers*, 43(3), pp. 341–359.

14. Oteyo, I.N., Marra, M., Kimani, S., Meuter, W.D. and Boix, E.G., 2021. A survey on mobile applications for smart agriculture: Making use of mobile software in modern farming. *SN Computer Science*, 2(4), p. 293.

15. Ciaburro, G. and Iannace, G., 2021. Machine learning-based algorithms to knowledge extraction from time series data: A review. *Data*, 6(6), p. 55.

16. Citoni, B., Fioranelli, F., Imran, M.A. and Abbasi, Q.H., 2019. Internet of Things and LoRaWAN-enabled future smart farming. *IEEE Internet of Things Magazine*, 2(4), pp. 14–19.

17. Gupta, M., Abdelsalam, M., Khorsandroo, S. and Mittal, S., 2020. Security and privacy in smart farming: Challenges and opportunities. *IEEE Access*, 8, pp. 34564–34584.

18. Sivasankari, A. and Gandhimathi, S., 2014. Wireless sensor based crop monitoring system for agriculture using Wi-Fi network dissertation. *International Journal of Computer Science and Information Technology Research*, 2(3), pp. 293–303.

19. Ferrández-Pastor, F.J., García-Chamizo, J.M., Nieto-Hidalgo, M., Mora-Pascual, J. and Mora-Martínez, J., 2016. Developing ubiquitous sensor network platform using internet of things: Application in precision agriculture. *Sensors*, 16(7), p. 1141.

20. Tang, Y., Dananjayan, S., Hou, C., Guo, Q., Luo, S. and He, Y., 2021. A survey on the 5G network and its impact on agriculture: Challenges and opportunities. *Computers and Electronics in Agriculture*, 180, p. 105895.

## Chapter 5

Harnessing AI for advanced crop management and decision support in agriculture

Almas Begum, K. Senthil, and S. Alex David

5.1 INTRODUCTION

The increasing unprecedented rate of population growth in the world means that the need for food is on the rise. To do so, agriculture must become more efficient and sustainable. Artificial intelligence (AI) has revolutionized the agriculture space by bringing innovative solutions to age-old problems. This chapter discusses how AI enables farmers to make informed decisions based on data, allocate resources more efficiently, and boost crop productivity.

The 21st-century dilemma confronting humanity is a unique and pressing one: we must provide an increasing world population, which is expected to exceed 10 billion people by 2050, with food [1]. The agriculture sector is under extreme pressure due to the massive population increase to deliver more food in human history [2]. The farming systems that have sustained us for centuries are no longer sufficient to satisfy the increasing demand for food, fiber, and bioenergy.

AI in agriculture has evolved from simplistic task automation to the integration of modern IoT. It was originally utilized for straightforward automation tasks, such as an irrigation system or a robotic milking machine. However, as technology drove AI's role to that of data-driven decision-making, the development of IoT sensors allowed farmers to access large swathes of data on soil health, crop conditions, and pest dangers, among other things. It can be used in a variety of contexts to apply a more specific application, notably in data analysis. Today, AI and IoT serve as key blowers in agriculture, reshaping farming techniques and increasing productivity through more sophisticated data analysis and predictive modeling.

Figure 5.1 shows the different ways AI has been utilized by agriculture over time. Early Automation: The timeline begins in the 1950s, the first attempts at automation in agriculture. Basic mechanization and elementary data collection methods characterized the early history of agricultural automation, as illustrated in [3]. With the development of computing power and data collection capabilities, this level progresses into the digital age 1990s–2000s. During this period, the initial AI applications such as crop modeling started

DOI: 10.1201/9781032618173-6

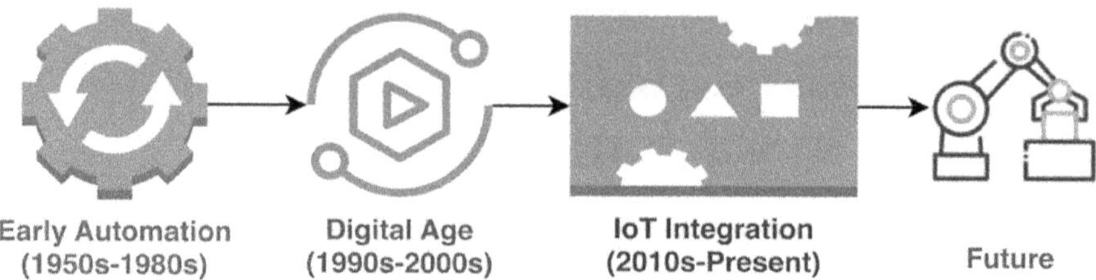

Figure 5.1 Evolution of artificial intelligence (AI) in agriculture.

to materialize. IoT Integration: The graph indicates significant alterations that occurred in the 2010s as a result of the integration of IoT into agriculture. Cloud computing, AI, and IoT sensors converge to allow real-time data collection and analysis. Current Situation: In the contemporary period, AI and IoT applications in agriculture have reached their high point. The timeline presents various cutting-edge application examples of AI, including precise farming, self-driving vehicles, and data-driven decision support systems. This timeline goes into the future to represent further possibilities, such as swarming robots, a statistical combine, and the broader use of AIs in agriculture.

5.1.1 AI's potential for agriculture

Artificial intelligence, or AI for short, encompasses a comprehensive array of technologies that enable machines to imitate the intellect of people. Machine learning, a subdiscipline of AI, has received a lot of attention in the agricultural sector. Algorithms may be educated to forecast or reach a conclusion based on evidence without any specific instructions. By learning from and adapting to data, the methods farmers use to manage their crops are becoming increasingly state-of-the-art. AI has many advantages for agriculture [4], including:

- **Use of Vast Amounts of Data:** The AI system processing is not limited to the data of one particular source but includes satellite imaging, weather sensors, soil sensors, and numerous other sources. A farmer using the data can analyze the best time to sow and water the crop as well as the correct time to harvest, which, in turn, maximizes production.
- **Precision Agriculture:** As a result of using AI, a farmer can give each field exactly what it needs and treat it in the way the field needs it. It allows minimizing the usage of resources like water, fertilizers, and pesticides; improves yield; and decreases environmental repercussions of farming. The most prominent examples of AI's potential in agriculture are drone imagery analysis and soil moisture prediction models. Drones with AI can provide insights into crop health and field conditions in real time, and predictive models enhanced with AI

can optimize irrigation schedules during droughts. Thereby, it significantly increases water efficiency and crop yield.

- **Picture Identification:** AI-powered picture identification can identify early stages of crop diseases and pests to target their management.
- **Early Disease Detection:** AI reduces the need for chemical treatments and crop losses.
- **Crop Yield Prediction:** AI algorithms can predict crop yields with almost perfect accuracy due to historical data evaluation and present conditions check. As a result, farmers and agribusinesses can plan marketing and logistics needs.
- **Resource Allocation Optimization:** AI can make recommendations about what inputs should be applied and when and where, such as water, fertilizers, and pesticides, reducing production costs and eliminating waste.

In this chapter, we delve into AI applications in agricultural areas. Particular emphasis will be placed on crop management. We will discuss the various AI methods from machine learning to deep learning that have made their way to the farm and are applied to data processing from IoT sensors. The real-life case studies and examples will show how AI is used for crop disease diagnosis, yield prediction, and resource optimization. Finally, the future AI opportunities and challenges in agriculture as well as the ethical and environmental concerns related to it will be discussed. While crop management will be the focus of this chapter, it is important to understand that AI's involvement in agriculture extends beyond it. From AI-driven autonomous farming equipment development and supply chain optimization to livestock management, with its direct impact on food production, crop management becomes a point of specific attention in terms of AI incorporation.

5.2 AGRICULTURE AND AI

5.2.1 The evolution of AI

A subfield of computer science focused on developing "intelligent" machines, AI has made significant strides in recent years. AI has, in particular, advanced in machine learning, natural language processing, and computer vision. However, the idea of AI has been floated for decades. While AI has revolutionized several sectors in recent years – mainly health care, banking, or transport – agriculture is not left out. The genesis of AI in agriculture is attributed to the early automation of farm activities. However, it has not thrived to this extent due to improved processing capabilities, data availability, and higher algorithmic intelligence. AI is now expected to transform agriculture by providing farmers and agronomists with data-driven advice and decision-making tools. AI boosts inputs used in agriculture by accurately predicting the required volume of water or nutrients, for instance, by crops, eliminating waste. AI models rely on sophisticated algorithms and historical patterns to calculate crop growth stages, soil moisture content, temperature patterns, and other factors. As a result, they can customize irrigation and fertilization plans to use resources efficiently and achieve the maximum possible returns.

5.2.2 A quick overview of AI in agriculture

There are several AI applications in agriculture which can be dejected from automating farming, agricultural machinery management, and crops and livestock management, as seen in Figure 5.2. AI employed in farming can help raise yields, eradicate waste, and do little harm to the world.

The use of AI aids in managing farms from crop handling to the health of the plants. AI allows farmers to make informed crop choices by providing them with immediate information on their soil's health, the status of the weather, and pest infestations on their plants. In addition, farmers will keep track of and manage their activities on their plants. This will be done via numerous software programs that may monitor and regulate a variety of plant treatment items. Furthermore, AI may be used in the following ways: monitoring the livestock – AI sensors and data analytics are used to keep an eye on the health and welfare of livestock by determining the species' feeding and medical care procedures; precision farming – AI allows farmers to apply resources such as water, fertilizer, and pesticides in the correct volumes necessary; automation of farm equipment – AI-enabled autonomous tractors and drones have the capability to do several jobs such as planting, harvesting, and pest eradication.

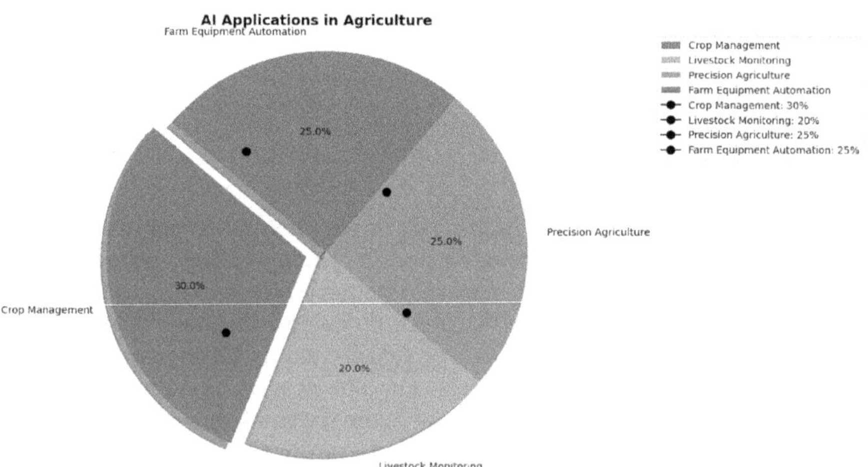

Figure 5.2 Application of artificial intelligence (AI) in agriculture.

5.2.3 The use of IoT in agriculture

Since the basis for the use of AI in agriculture is the Internet of Things (IoT), we have multiple IoT devices, including sensors and actuators, operating in the field to provide the necessary information on many environmental parameters, including soil moisture, soil temperature, air humidity, and insect activity. AI-based cloud platforms obtain this data and analysis and produce optimal controlling or decision recommendations.

5.3 IoT SENSORS AND DATA GATHERING

5.3.1 IoT sensor types for agriculture

The Internet of Things, referred to as IoT, is the ecosystem of connected devices that gather and share information across the Internet. In AgriTech, IoT refers to the application of sensors, drones, and other connected devices to collect real-time information on various elements regarding the farm – soil moisture, temperature difference, humidity, crop health, and more. This information is thereafter transmitted to a central system from where it is analyzed to enable data-driven sound decisions, best practices in resource allocation, and improved management practices overall.

IoT sensors for agriculture come in a variety of shapes and sizes, each one intended to gather a particular kind of data [5]. Various sensors used in agriculture are shown in Figure 5.3. IoT sensors are frequently utilized in agriculture in the following ways:

- Soil moisture sensors help farmers optimize irrigation and minimize over- or under-watering by measuring the amount of moisture in the soil.
- Weather sensors in IoT weather stations measure temperature, humidity, wind speed, and precipitation, which are critical in data gathering for crop management.
- Pest monitoring sensors, which measure the presence of pests in the field, allow farmers to take action as soon as possible.
- Crop health sensors detect the health of crops by taking visual data such as leaf color and texture using cameras and image recognition software.
- GPS and geospatial sensors, which track exact locations, aid in precise farming mapping and analysis of farms geospatially.
- Sensors for livestock, IoT sensors, will also monitor livestock behaviors and general health. Heart rate, body temperature, and feeding habits are examples of such monitoring aspects.

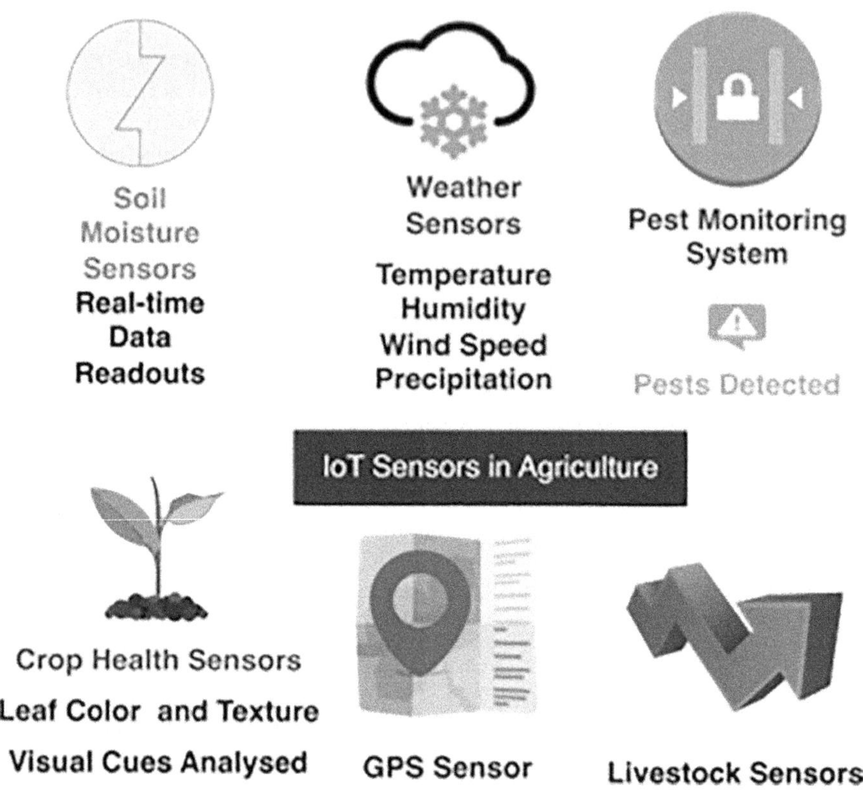

Figure 5.3 Internet of Things (IoT) sensors in agriculture.

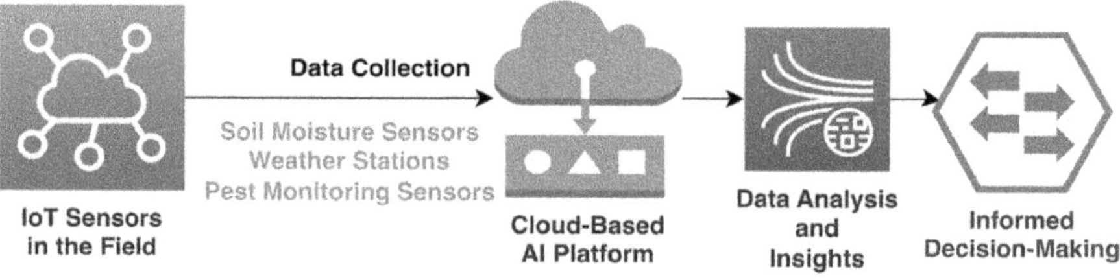

Figure 5.4 Data flow in the Internet of Things (IoT)-enabled agriculture.

5.3.2 Data management and collection

Although data collection using data IoT sensors is only a small portion of it, the sensors produce large amounts of data, which need to be stored and processed to give useful insights. Sensors collect geospatial data, pictures, and time-series information as presented in Figure 5.4. This data is easily managed and analyzed by cloud-based platforms and data analytics tools that are common among farmers and agricultural professionals. The platforms are able to preprocess or analyze data in real time and produce actionable insights. It is also possible to use this data to train machine learning and deep learning models, which can also be used for predictions or suggestions.

5.3.3 Data collection challenges

IoT sensors have numerous benefits for data collecting, but they also present issues that must be resolved:

- **Data Quality:** Data gathered by IoT sensors may also be unreliable and inaccurate due to sensor calibration, environmental conditions, and sensor failure. Quality data is essential for smart decision-making.
- **Data Security:** Data on crop yields, pest attacks, and irrigation timetables are important farming information and hence beneficial. Such data must be kept safe from unauthorized access and online threats. Encryption for secure data transmission and storage, additional access control, and standardized data formats for interoperability are all solutions. It is critical to enhance the connection using secure protocols and keep the data up to date. Even though encryption is perfect for data integrity, cybercriminals can intercept encrypted data. Security measures, such as encryption, surveillance, and training, and proper planning audits can be used to aim and explain the pitfalls.
- **Data Fusion:** Many farms have numerous sensors from many manufacturers, all of which have various communication formats and data transmission technologies. It may be challenging to integrate data from additional sources.
- **Connectivity Concerns:** Low-quality Internet connections in some locations might make it difficult for IoT sensors to transmit real-time data to cloud-based platforms.

5.4 CROP MANAGEMENT MODELS USING MACHINE LEARNING

5.4.1 Supervised crop management learning

A machine learning paradigm, known as supervised learning, uses models that have been trained on labeled data to generate predictions or categorize data. Supervised learning is frequently used in crop management for tasks including weed identification, yield prediction, and crop disease detection [6].

As seen above, machine learning can help to detect crop diseases early. Similar to crop diseases, supervised machine learning models can be trained on labeled photos of plants taken at different stages of growth with and without particular diseases or visual indicators. Reading that visual indication may be loss malformation or leaf discolor. Other uses of machine learning in agriculture include crop yield prediction; supervised machine learning models can be trained with historical agricultural yield data, comprising elements such as weather, soil quality, and irrigation. Identifying weeds is essential because they compete with crops for vital nutrients, leading to reduced yields. Supervised machine learning models may be trained to recognize and differentiate specific types of weeds from the crops themselves to help get rid of the issue.

In another supervised learning, crop management can apply a model that is trained using photos to identify and classify tomato blight accurately. The model is trained to identify the specific symptoms of the blight by the use of labeled data of a healthy tomato and affected tomato plant. Once the blight is identified, it can be addressed immediately, preventing the spread of the disease and minimizing the loss of farm produce through low yields. The prediction from the model developed through supervised learning makes it possible for farmers to act proactively to prevent losses and boost their farm produce.

5.4.2 Unsupervised pattern recognition learning

Unsupervised learning is another machine learning technique used for clustering and pattern detection in crop management. K-means clustering and hierarchical clustering are two unsupervised learning methods that can be used to find hidden patterns in agricultural data [7].

Below are few applications of unsupervised learning for agricultural data:

- **Classification of soil types:** Using the unsupervised learning method, field areas with similar soil properties can be united into specific groups, which will ensure more efficient soil treatment and fertilizer distribution.
- **Planning the crop rotation:** Unsupervised learning allows preparing an optimal crop rotation schedule for healthy soil by forming groups of fields according to previous crop yields and soil characteristics, among which crop rotating will bring more profit.
- **Investigating pests' activity:** Unsupervised learning helps to detect patterns in pests' data and make decisions on more efficient pest control management systems and pest breeding forecasts.

5.4.3 Reinforcement learning in precision agriculture

An AI paradigm known as reinforcement learning teaches agents how to operate in a given environment to maximize a reward signal [8]. Reinforcement learning can be used in precision agriculture to improve resource management and machinery control.

- **Irrigation Control:** Reinforcement learning models can learn how to properly control irrigation systems by considering factors such as crop growth stage, weather predictions, and soil moisture levels in harvesting foodstuff while conserving water.
- **Autonomous Farming Equipment:** Reinforcement learning theory is used to enable drones and autonomous tractors to fly and navigate fields and decide when they can plant seeds and harvest crops wisely and also recognize obstacles. For example, intelligent pest management systems use drones or robots to focus on specific areas where insects or pathogens are discovered rather than chemical sprays with wide effectiveness.

5.5 DEEP LEARNING CROP MANAGEMENT ALGORITHMS

5.5.1 Image analysis using convolutional neural networks

Deep learning helps computers decipher trends in large datasets. It aids in determining crop diseases from images since the technology may examine visual data and detect abnormal growth or issues before they impede ability. Convolutional neural networks (CNNs) have been proven to be quite effective for agricultural picture analysis activity. These deep learning algorithms can analyze crop image data and identify abnormalities, sicknesses, and trends.

- **Detection of Diseases:** CNNs use images of plant leaves to identify and diagnose diseases based on visual signs, and farmers can detect diseases early enough.
- **Identification of Weeds:** CNNs can identify between crops and weeds in pictures, making it easier to manage weeds.
- **Monitoring of Crop Growth:** CNNs can track the growth of plants by checking several photos of crops with time to understand the status of the crops.

5.5.2 Time-series data with recurrent neural networks

Recurrent neural networks (RNNs) are useful in agriculture for tasks like weather forecasting, irrigation scheduling, and insect behavior monitoring because they excel at processing time-series data [9].

Recurrent Neural Networks (RNNs) leverage both current conditions and historical data to predict short-term and long-term weather patterns, enabling farmers to effectively plan irrigation and other agricultural activities. By analyzing patterns in soil moisture data, RNNs can develop and recommend optimal irrigation schedules, which help conserve water while ensuring the preservation and health of crops. Additionally, RNNs play a crucial role in pest management by monitoring time-dependent information on pest activity. This allows RNNs to predict potential pest outbreaks, assisting farmers in determining when and where to implement targeted pest control measures. Overall, the application of RNNs in agriculture enhances resource optimization, reduces waste, and contributes to sustainable farming practices.

5.5.3 Agriculture and generative adversarial networks

In agriculture, generative adversarial networks (GANs) are utilized for data generation, augmentation, and enhancement activities [10].

- GANs could generate artificial pictures of pests and crops. These phony photographs may be used to increase the number of photos in a machine learning dataset. Noise reduction, resolution boosting, and color imbalance adjustment may improve the photos in image recognition tasks. This is really useful for detecting raindrops and bugs. Moreover, artificial pests created using this technique could also serve as negative training samples for machine learning.

5.6 DETECTION OF CROP DISEASES

5.6.1 Early detection via AI

GANs could generate artificial pictures of pests and crops. These phony photographs may be used to increase the number of photos in a machine learning dataset. Noise reduction, resolution boosting, and color imbalance adjustment may improve the photos in image recognition tasks. This is really useful for detecting raindrops and bugs. Moreover, artificial pests created using this technique could serve as negative training samples for machine learning.

5.6.2 Case study: AI-based plant disease detection

Let's look at a case study from the real world to show how AI is revolutionizing crop disease detection.

5.6.2.1 AI-powered disease detection in tomato crops case study

The tomato yellow leaf curl virus, late blight, and early blight are among the illnesses that can affect tomatoes. Early disease detection is essential for decreasing crop losses and is a requirement for chemical treatments.

- **Solution:** An AI-based disease detection system for tomato crops was created by a research team. To take pictures of plant leaves, IoT sensors with cameras were deployed in tomato fields. The deep learning model, a CNN that was trained exclusively for disease identification, was then fed these photos.
- **Results:** The AI system was quite successful at spotting diseases in tomato crops. Even in the early phases when visual clues were subtle, it could accurately identify illness signs. Due to the farmers' ability to take focused action, such as modifying irrigation and using the proper treatments, crop losses and the need for pesticides were reduced. Sample input–output has been shown below, and the detection distribution is shown in Figure 5.5.
- **Sample Result:**

1. **Input:** Taking a picture of a tomato leaf
   Output: Disease detected: late blight
   Action to Be Taken: Taking action: modifying irrigation and using proper treatments
   Outcome: Early detection and proper management of late blight can increase crop yield by up to 15% as well as AI recommendations

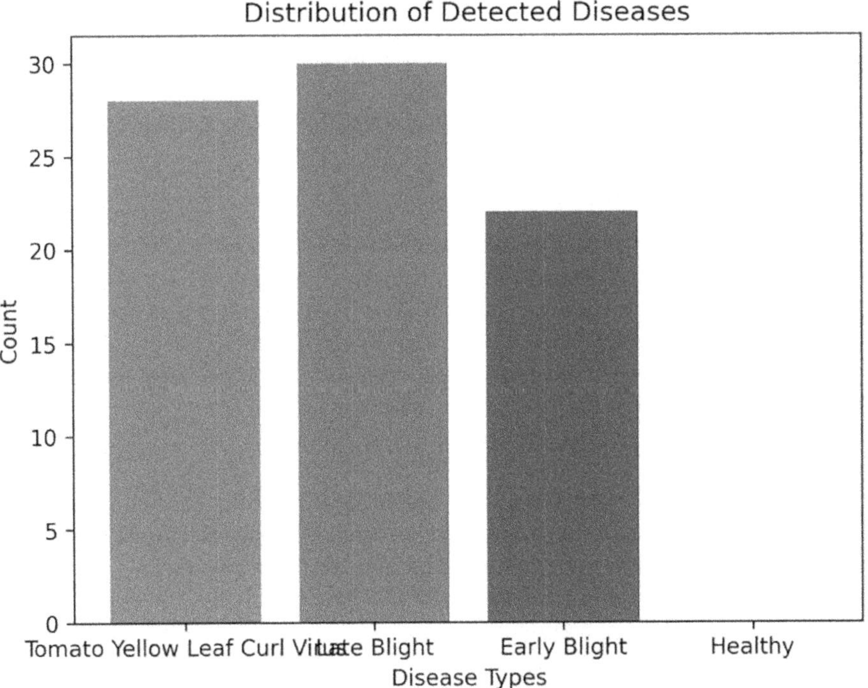

Figure 5.5 Distribution of detected diseases in the tomato leaf.

reduce pesticide applications, resulting in a 20% reduction in pesticide use compared to traditional methods.

2. **Input:** Taking a picture of a tomato leaf
   Output: Disease detected: early blight
   Action to Be Taken: Modifying irrigation and using proper treatments
   Outcome: Detecting early blight allows farmers to implement timely measures, resulting in a 10% increase in crop yield.

3. **Input:** Taking a picture of a tomato leaf
   Output: Disease detected: tomato yellow leaf curl virus
   Action to Be Taken: Taking action: modifying irrigation and using proper treatments
   Outcome: The detection of tomato yellow leaf curl virus helps prevent yield losses of up to 30% by enabling targeted management practices. By implementing AI-recommended strategies, farmers experience a 25% reduction in pesticide use and achieve good crop yield.

4. **Input:** Taking a picture of a tomato leaf
   Output: No disease detected. No action required.
   Outcome: AI-powered disease detection provides a 5% reduction in overall pesticide use and conserves resources while maintaining crop productivity.

Crop yields increased as a result of early disease identification, increasing farmer profitability.

Lessening the usage of chemical treatments decreased manufacturing costs and the impact on the environment. Throughout the growing season, the AI system's continual monitoring allowed for prompt interventions.

5.6.2.2 Benefits and difficulties

The use of AI for crop disease detection has the following advantages:

AI systems can spot diseases before they spread, allowing for quick action and minimizing crop losses. AI can identify between various diseases and provide exact information on the kind and severity of the ailment. AI-based systems can keep an eye on crops constantly, offering real-time insights and obviating the need for manual inspections.

However, there are obstacles to take into account:

- **Data Quality:** The efficacy of disease detection models is largely dependent on the quality of the training data, which must contain a variety of samples of both healthy and diseased plants.
- **Scalability:** Setting up large-scale AI-based illness detection systems can be expensive, requiring the installation of lots of IoT sensors and computer power.
- **Accessibility:** Adopting AI-based solutions may be difficult for small-scale farmers with limited resources.

5.7 YIELD PREDICTION

5.7.1 Factors affecting crop yield

Crop yield forecasting is an essential component of crop management since it aids in planning harvests, resource allocation, and decision-making for farmers [11]. Crop yield is affected by a number of factors, including:

- **Weather:** Crop growth and yield are directly impacted by temperature, precipitation, and sunlight.
- **Soil Quality:** Crop health and productivity are influenced by the composition of the soil, which includes nutrient levels and pH.
- **Pest and Disease Pressure:** If not properly controlled, the presence of pests and diseases can lower production.
- **Water Availability and Irrigation:** Proper and timely irrigation is crucial for crop growth and output.
- **Crop Variety:** Under particular circumstances, many crop types have variable production potentials.

5.7.2 AI-based yield prediction models

To come up with a prediction of farm produce, yield prediction models which are AI-based use data for both past and present years based on the weather pattern, soil quality, pests, and diseases among other factors. In this, regression models, decision trees, and neural networks are some of the machine learning techniques mostly used.

The procedure usually entails:

- **Data Collection:** Compiling information from historical records, weather stations, and IoT sensors.
- **Feature Engineering:** The process of choosing and processing pertinent features (variables) that impact crop output.
- **Model Training:** To anticipate agricultural yields, a machine learning model is trained using historical data.
- **Model Validation:** Analyzing validation datasets to judge the model's generalizability and correctness.

Making forecasts for current or next growing seasons based on input data is known as yield prediction.

The foremost method to ensure the accuracy of yield prediction models is to validate AI yield predictions against actual harvest outcomes through multiple seasons. Actual results are the ultimate measure, despite the fact that AI models forecast harvests by employing past harvests, weather data, soil conditions, and so on. Differences between expected and actual yields enable one to compare and contrast how well or badly a model works. The model can be demonstrated to be reliable or unreliable by consistently predicting right or inaccurately across seasons. Although the former is critical in boosting reliance on one's model's ability to forecast, the latter stimulates iterative enhancements. Ultimately, this process's constant loop improves decision-making in agriculture and safeguards sustainable farming processes.

5.7.3 Case study: AI prediction of crop yield

Let's look at a case study that exemplifies the application of AI to crop yield forecasting.

5.7.3.1 AI-Based Crop Yield Prediction for Wheat Case Study

- **Background:** Due to variables including weather and soil quality, the production of wheat, a staple crop, can vary significantly. Farmers must be able to anticipate yields accurately to plan harvests and use resources as efficiently as possible.
- **Solution:** A team of academics created a model for predicting wheat crop yields using AI. Over a number of years, they gathered information on the weather, the state of the soil, and past crop yields from various farms. They developed a gradient-boosting approach using this data to train a machine learning model to forecast wheat yields depending on input factors.
- **Results:** When predicting wheat yields, the AI-based yield prediction model showed outstanding accuracy. Based on weather and soil data from the beginning of the season, it might offer yield projections for the next year. Utilizing this approach, farmers might modify their planting and resource allocation plans to increase output potential. Sample data is shown below, and prediction has been plotted in Figure 5.6.

Sample Data:
Input:
Years: [1 2 3 4 5 6 7 8 9 10]
Weather Data (Temperature): [27.12172794 23.94121793 25.35914124 23.63550767 27.35249492 23.29094788 24.93703426 28.47371169 25.65372595 26.86405235]
Soil Quality Data: [53.76405235 57.49765193 47.81075841 58.52783919 49.9692348 45.84690491 53.53553633 43.01525656 49.55376644 59.30581002]
Output:
Predicted Crop Yields: [6456.18639712 6153.75949762 5981.08463227 6191.3138481 6710.24435824 5702.97082411 6228.03444271 5798.94456792 6531.52539271 6470.86855825]
Benefits:

- **Better Resource Allocation:** Based on anticipated yields, farmers might use fertilizer and irrigation more efficiently, cutting costs and waste.

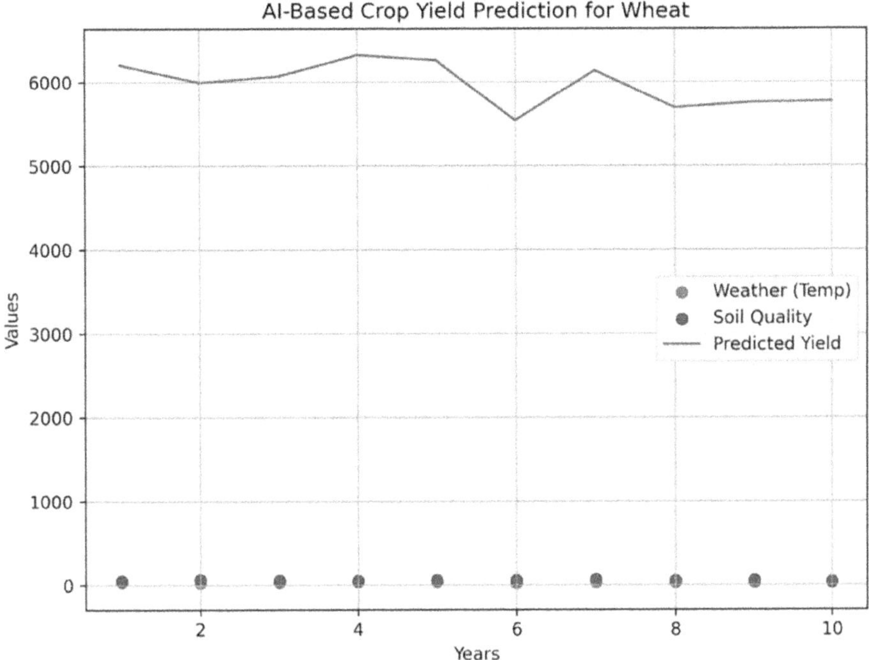

Figure 5.6 Wheat yield prediction using artificial intelligence (AI).

- **Risk Reduction:** By investigating alternate crops or tactics, farmers were able to get ready for probable produce deficits due to early yield estimates.
- **Data-Driven Decision-Making:** The AI model offered data-driven insights that could be put to use by farmers to manage their crops.

5.8 AGRICULTURE RESOURCE OPTIMIZATION

5.8.1 Management of water

Water is a scarce resource in agriculture, and effective water management is necessary for sustainable farming over time. Because AI provides up-to-date information on soil moisture and weather, it plays a crucial role in water usage maximization. The following are some of the areas where AI contributes to water use maximization in agriculture: soil moisture sensing – using IoT sensors in the field, moisture amounts are being recorded at various depths in the soil. AI algorithms analyze this data to determine when and how much irrigation is essential to prevent overwatering and water wastage. Irrigation based on weather – past weather data and weather predictions are used by AI systems to adjust irrigation timetables. As a result, in any given period, crops receive water based on the weather, either real or anticipated. Drought prediction – following past weather patterns, machine learning models predict drought, assisting farmers in preparing for the disaster and implementing water-saving measures.

Also, according to economic analysis, AI technology indeed increases agricultural sustainability by maximizing resource allocation and cutting operational costs, but it also generates tangible cash gains. AI-driven optimization of water management in agriculture: first, it results in significant cost savings, along with water usage reduction. Using AI for water optimization via weather-based irrigation and soil moisture acquisition, farmers cut water usage and expenses. However, when it comes to drought prediction models of AI implementation, it becomes possible to plan in advance and thus mitigate the impacts, ensure no crop is lost, and, as a result, not lose money from the lost crop.

5.8.2 Application of fertilizers and pesticides

Moreover, both fertilizers and pesticides need to be better managed to reduce their negative impact on the environment – for that matter, the cost of manufacture, as well. An AI-driven algorithm has a recommendation based on current data and historical experience. Nutrient management involves an AI analysis of crop development and soil heat using algorithms to identify fertilizers and the best timings to use them to reduce excessive fertilizer contamination. AI can predict pest outbreaks by analyzing images and sensor data, ensuring that the target pesticide application occurs solely at necessary times in areas of infestation to rely only on the required sacrifices. Integrated Pest Management (IPM) methods are advantageous and complement AI systems because they incorporate organic, cultural, chemical, and biological approaches to pest control, resulting in reduced use of chemical pesticides.

5.8.3 Case study: AI-powered resource optimization

Let's look at a case study that demonstrates how AI might improve agricultural resource management:

5.8.3.1 AI-based resource optimization for precision farming case study

- Background: The management of water and fertilizer presented difficulties for a large-scale farm. Traditional irrigation techniques wasted water because they were ineffective, and fertilizer applications were not made to account for particular field circumstances.

- **Solution:** The farm adopted a precision farming approach based on AI. Real-time soil moisture, temperature, and nutrient levels were tracked by IoT sensors. To analyze this data and offer suggestions for the application of fertilizers and irrigation, machine learning models were trained.
- **Results:** Sample input year, water-saving, and yield-based analysis is shown below, and optimization has been shown in Figure 5.7.

Input:

Years: [2010, 2011, 2012, 2013, 2014, 2015, 2016, 2017, 2018, 2019]
Water Savings (%): [0, 10, 15, 20, 25, 28, 30, 32, 35, 38]
Increased Yields (%): [0, 3, 6, 8, 10, 12, 15, 18, 20, 22]
Lower Costs (%): [0, 5, 10, 12, 15, 18, 20, 22, 25, 28]

Resource management significantly improved thanks to the AI-driven system:

- **Water Savings:** The farm implemented precision irrigation based on soil moisture levels and weather forecasts, reducing water usage by 30%.
- **Increased Yields:** Proper nutrient treatment and timely irrigation led to a 15% increase in crop yields.
- **Lower Costs:** The farm spent less on water and fertilizer, which increased its profitability.

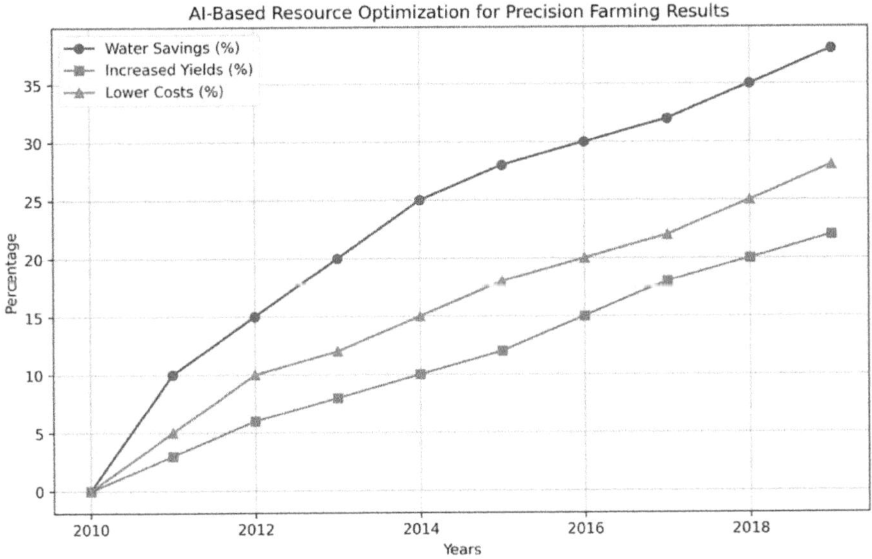

Figure 5.7 Resource optimization with artificial intelligence (AI) for farming.

5.8.3.2 Benefits

- **Environmental Impact:** The farm's decreased use of water and chemicals resulted in less runoff and pollution, which was good for the environment.
- **Scalability:** By expanding the AI system's coverage of the farm's broader areas, resource management will be further improved.
- **Data-Driven Choices:** The farm made data-driven choices that increased sustainability and overall efficiency.

5.9 USING AI TO IMPROVE FARMING EQUIPMENT

5.9.1 Autonomous farm equipment

Integration of AI with farming equipment is altering the agricultural environment. Various activities can be performed more efficiently with less labor using autonomous tractors, drones, and robotic devices with the assistance of AI. Autonomous tractors, equipped with AI technology, can drive, plow, plant, and harvest crops without human intervention. These machines employ sensors, GPS, and computer vision to operate precisely. Drones are used in agriculture, and they can inspect the health of crops, find pests and illnesses, and administer treatments. They provide a bird's-eye view of the situation to quickly take in and analyze the data. Robotic harvesters can harvest crops including fruits and vegetables with pinpoint precision, reducing yield loss while boosting quality.

Some of the safety measures or protocols that can be added while using autonomous farm equipment to take actual safety precautions include the following:

1. **Training Operators:** Make sure that the operators using the autonomous farm equipment are well trained on how to operate it safely.
2. **Safety Sensors:** The safety sensors used to detect objects or human beings or any hazards available in the environment are installed in equipment automatically at emergency stops.
3. **Emergency Stop Mechanisms:** The emergency stop buttons or switch should be fitted in the equipment or installations, which allows the operators to stop using the equipment quickly if any kind of safety hazard arises.
4. **Emergency Response Mechanisms:** The emergency response plan for responding to accidents, injuries, or injecting equipment malfunctions is designed and developed.

5.9.2 AI-powered precision agriculture

By integrating AI into farming equipment, farmers can adopt an approach that fits current field conditions, known as precision agriculture. AI systems assess data from farm gear sensors and cameras to create decisions in real time. AI systems can save resources with the aid of the following techniques: Variable rate application – AI adjusts the rate at which fertilizers, pesticides, and water are spread depending on soil circumstances and crop conditions; weed control – autonomous weeding robots equipped with AI can identify and eliminate weeds without the use of chemicals, significantly lowering herbicide use; AI-powered drones may photograph fields and identify troubling regions, such as disease or nutrient insufficiency, allowing for targeted treatments.

5.9.3 Upcoming farm equipment trends

Some of the potential uses of AI in farm machinery include using robot swarms in planting and harvesting. These swarms, comprising small networked machines, could work collaboratively to deliver these operations more efficiently. Another potential use is the use of combiners with AI power. In the future, combiners could sort and process crops after harvesting without human input to reduce post-harvest losses. Additionally, an AI algorithm can optimize the energy use of machines. This can reduce the amount of fuel used and minimize environmental repercussions. Lastly, farmers can monitor and control machines with AI functions remotely using a computer or cellphone.

5.10 ENVIRONMENTAL AND ETHICAL ISSUES: FUTURE SCOPE

5.10.1 Agriculture's sustainability

Another factor where AI technologies and agriculture meet is sustainability. Sustainable farming is a type of farming that tries to maintain productivity while reducing the negative impact it has on the environment. AI, by optimizing the use of land, reducing waste of various resources, and promoting environmentally friendly farming practices, can be of assistance in the field of sustainability:

- **Reduced Need for Chemical Pesticides and Herbicides:** AI-driven systems for pest and disease detection minimize the reliance on chemical treatments, promoting environmentally friendly farming practices.
- **The Efficient Use of Resources:** AI-powered precision agriculture can dramatically reduce resource consumption by optimizing water use and the use of fertilizers.
- **Soil Health:** AI can work to optimize sustainable farming practices by improving soil health through crop rotation and no-till.

5.10.2 Ethical issues in agriculture driven by AI

The application of AI in agriculture raises several ethical concerns that must be resolved. First, the data on the farm, including crop yield and farming methods, is both delicate and valuable; security and privacy must be guaranteed. It may seem difficult to impossible for fewer wealthy or smaller farms to access and use AI technology to perhaps exacerbate disparities in agriculture. As tasks on the farm become more automated, employees may lose their jobs. People must be retrained and supported in their transition to jobs. While AI has the ability to support sustainability in many areas, it must be carefully applied to prevent unforeseen harm to the environment.

When implementing AI resources for small farmers in community-based settings, it is essential to address complex ethical considerations to ensure fair access and equitable distribution of benefits. Firstly, the implementation should ensure that more farmers gain access to the technology. Additionally, transparency is crucial; clear explanations of how the AI systems work, their limitations, potential biases, and their impact on farming practices must be provided. Farmers should have greater control over their data, supported by strong protective laws to safeguard their information. Furthermore, data ownership should be transferred to the farmers, ensuring their privacy is maintained.

It must be stressed that farmers should be engaged in the technology development process to resolve such issues plaguing their community. Affordability and accessibility should be budget constraints, as the community is struggling, and all levels of funding and investments must be considered. Local empowerment enables enhanced cooperation and information sharing. It also allows communities to make decisions regarding technology implementation and use based on their beliefs. Advantages can be distributed more fairly so that everyone involved benefits financially regardless of where they are located or their relative financial status. Long-term sustainability, with a focus on the technology's impact on regional ecosystems, economies, and social interactions, refers to approaches to address: environmental stewardship, preservation of cultural heritage, and resilience of communities.

5.10.3 Frameworks for rules and policies

Governments and agricultural organizations are creating rules and regulatory frameworks for AI in agriculture to address ethical and environmental problems. These guidelines are designed to guarantee ethical AI use, data security, and equitable access to technology.

5.10.4 Emerging technologies

The future of AI in crop management seems to have several promising technologies:

- **Quantum Computing:** Quantum computing can accelerate complex AI calculations, opening the possibility of more sophisticated models and faster decision-making.
- **Edge AI:** Edge computing and AI can run real-time processing of data from any number of IoT devices, reducing the latency and increasing efficiency.
- **Explainable AI:** Advances in this direction would enable farmers to better understand AI's recommendations and predictions, thus becoming more self-confident. Widespread acceptance herewith, there are challenges to AI in agriculture: most areas of application still face challenges.

Building AI infrastructure and implementation may be expensive for small farmers.

- **Data Accessibility and Quality:** AI requires solid, diverse data, which might be hard to access and ensure its quality in out-of-date or rural locations.
- **Education and Training:** Farmers and agricultural professionals should be trained to use AI instruments and understand AI-provided insights. It is hard to overstate the importance of AI for food security. AI-driven crop management could oversee and ensure stable food supplies, as well as reduce food spoilage and adjust to changing environmental conditions corresponding to the world population rate.

5.11 CONCLUSION

The integration of AI and the IoT has sparked a new era of experimentation in agriculture, one that has both fulfilled and exceeded its initial promises of innovation and sustainability. As it has grown from its humble beginnings to its current technical apex, AI has become a truly transformational power that can provide farmers and agronomists with data-driven answers. IoT sensors and AI-driven farming systems deliver real-time data on critical issues like soil health, weather conditions, and livestock well-being. Production is elevated, waste is diminished, and the impact on the environment is lowered merely due to the accuracy of the decisions enabled by this information abundance.

The significance of AI in disease diagnosis and yield prediction was emphasized in this chapter. As for yield prediction, AI can benefit by optimizing resources and reducing risks. In turn, disease early identification saves crops, reduces the environment's harm, or boosts yield. The optimization of resources with the help of AI has the potential to make farming sustainable and efficient, especially in water management and use of fertilizing substances and pesticides. Farming machines will be integrated with AI, including tractors with auto control and drones; in this regard, machines will act on their own, allowing for greater production and eliminating human power.

From an environmental and ethical perspective, sustainability and the proper use of AI in agriculture are undoubtedly issues that must be met. It is essential to ensure that the impact of AI in agriculture on global food security, environmental sustainability, and farmers is positive through stakeholder experience sharing, scientific attempts, and strict regulatory laws. The collaboration of AI and the IoT on the issues of production and sustainability afflicting agriculture appears to be the ideal solution. Artificial Intelligence of Things, which seamlessly integrates grower data, streamlines analysis, and uses AI in new, predictive ways, disrupts traditional agriculture entirely.

ACKNOWLEDGMENTS

I extend my sincere appreciation to my team members whose expertise has profoundly influenced "AI Applications in Decision Support and Crop Management". Special acknowledgment to the editors for their editorial refinements. This chapter benefits from collective contributions from and discussions with colleagues, enriching its content. I also thank my family for their understanding and encouragement during this endeavor. This work represents a shared commitment to advancing agriculture through AI applications. I also declare that this work is not supported by any funding agency.

REFERENCES

1. Searchinger, T., Waite, R., Hanson, C., Ranganathan, J., Dumas, P., Matthews, E. and Klirs, C., 2019. Creating a sustainable food future: A menu of solutions to feed nearly 10 billion people by 2050. Final report.
2. Boserup, E., 2014. *The Conditions of Agricultural Growth: The Economics of Agrarian Change under Population Pressure.* England, UK: Routledge.
3. Jha, K., Doshi, A., Patel, P. and Shah, M., 2019. A comprehensive review on automation in agriculture using artificial intelligence. *Artificial Intelligence in Agriculture*, 2, pp. 1–12.
4. Smith, M.J., 2018. Getting value from artificial intelligence in agriculture. *Animal Production Science*, 60(1), pp. 46–54.
5. Rathee, G., Sharma, A., Saini, H., Kumar, R. and Iqbal, R., 2020. A hybrid framework for multimedia data processing in IoT-healthcare using blockchain technology. *Multimedia Tools and Applications*, 79(15–16), pp. 9711–9733.
6. Akulwar, P., 2020. A recommended system for crop disease detection and yield prediction using machine learning approach. In Sachi Nandan Mohanty, Jyotir Moy Chatterjee, Sarika Jain, Ahmed A. Elngar, and Priya Gupta,

Recommender System with Machine Learning and Artificial Intelligence: Practical Tools and Applications in Medical, Agricultural and Other Industries (pp. 141–163). Wiley Online Library.

7. Alex David, S., Varsha, V., Ravali, Y. and Naga Amrutha Saranya, N., 2022. Comparative analysis of diabetes prediction using machine learning. In G. Ranganathan, Xavier Fernando, and Selwyn Piramuthu *Soft Computing for Security Applications: Proceedings of ICSCS 2022* (pp. 155–163). Singapore: Springer Nature Singapore.

8. Malik, A. and Tuckfield, B., 2019. *Applied Unsupervised Learning with R: Uncover Hidden Relationships and Patterns With K-Means Clustering, Hierarchical Clustering, and PCA.* Birmingham, UK: Packt Publishing Ltd.

9. Naranjo-Torres, J., Mora, M., Hernández-García, R., Barrientos, R.J., Fredes, C. and Valenzuela, A., 2020. A review of convolutional neural network applied to fruit image processing. *Applied Sciences*, 10(10), p. 3443.

10. Martinelli, F., Scalenghe, R., Davino, S., Panno, S., Scuderi, G., Ruisi, P., Villa, P., Stroppiana, D., Boschetti, M., Goulart, L.R. and Davis, C.E., 2015. Advanced methods of plant disease detection. A review. *Agronomy for Sustainable Development*, 35, pp. 1–25.

11. Nova, K., 2023. AI-enabled water management systems: An analysi of system components and interdependencies for water conservation. *Eigenpub Review of Science and Technology*, 7(1), pp. 105–124.

Applications and future directions

Part 2 shifts from foundational concepts to real-world applications, demonstrating the diverse ways AIoT is reshaping industries such as agriculture, healthcare, and cultural domains. It presents case studies on sustainable farming and medical innovations, highlighting the practical benefits of AIoT in diagnostics, personalized patient care, and chronic healthcare management. Additionally, the inclusion of a cultural application, with AI identifying Carnatic ragas, showcases the adaptability of AIoT across non-traditional sectors. This part also touches on future directions, especially in healthcare monitoring and diagnostics, reinforcing this book's exploration of AIoT's vast potential for innovation and improvement in everyday life.

DOI: 10.1201/9781032618173-7

Advancing smart agriculture through AIoT integration for sustainable farming solutions

Supriya Mallanayakanakatte Shankaregowda,
Rachana Reddy Koppala, and Rakshita Bhardwaj

6.1 INTRODUCTION

Smart agriculture is an innovative approach to farming; it can be significantly empowered by integrating it with Artificial Intelligence of Things (AIoT). This fusion of artificial intelligence (AI) and Internet of Things (IoT) technologies with agriculture can bring a noticeable transformation in traditional agricultural practices. It can also permit farmers to make decisions based on the data collected by AIoT and raise their crop production. In every domain, challenges are unavoidable; when it comes to agriculture, there are many obstacles, but one of the serious issues lies in its inability to sufficiently feed all those who are in need [1]. Since 1800, the global population has risen nearly 20 times [2]. Statistically, some reports suggest that global food production is expected to increase from 35% to 56% between 2010 and 2050, appealing the corresponding increase in food production. This obligation clashes with a gradual reduction in irrigation land due to ongoing developmental initiatives. Every technology has its unique range of applications. AIoT, for instance, has many applications; one among them is its ability to reduce energy wastage and ensure efficient energy maintenance. It also plays a powerful role in the financial sector, enabling the detection of fraud, risk management, and providing personalized banking experience.

Among various technologies, the integration of AIoT with agriculture is motivated by several factors. Chief among them is its ability to reduce expenses through efficient resource utilization and overall enhancement of agricultural efficiency. This smart integration also ensures sustainable and eco-friendly practices, reducing waste and promoting soil health. Moreover, AIoT empowers farmers with real-time insights and aids in decision-making. By integrating AIoT with agriculture, we can modernize and meet the demands of the growing population. Combining AIoT with agriculture presents both opportunities and challenges when it comes to livestock; AIoT can strengthen livestock monitoring and improve animal health and behavior. The challenge lies in the smooth integration of these complex technologies and addressing data security, privacy issues, and the need for such potential technology in agriculture.

DOI: 10.1201/9781032618173-8

Timely support is the key to preventing interruption caused by AIoT device maintenance. Additionally, adverse weather conditions can significantly disturb the performance and efficiency of these AIoT devices in agricultural operations. Raising farmers' awareness involves collaborating with agricultural universities and research institutes to develop and demonstrate prototypes. Circulating informative brochures that highlight AIoT's role in building up productivity, along with many other effective methods, is highlighted in this chapter.

This chapter's structure can be categorized into two main sections: the initial theoretical segment followed by the research segment. Section 6.5 delivers a practical implementation of AIoT in agriculture. Section 6.6 addresses the challenges associated with implementing AIoT in agriculture. Section 6.7 emphasizes improving awareness among farmers. Finally, Section 6.8 concludes this chapter [3].

6.2 CHALLENGES IN AGRICULTURE

The agriculture sector encounters multiple obstacles; pest attacks are one of the significant problems because if they are left untreated they destroy crops. Treating them with synthetic pesticides may lead to pollution, soil erosion, and a decrease in crop quality. There are high chances of these chemicals entering into food chain. Pests target different parts of the plants. Detector-based identification of affected parts can help in early detection.

In addition to pest-related challenges, agriculture is impacted by climatic change. Climate is not something which is under human control, but human activities will influence and affect climatic changes. Human activities, including burning fossil fuels, deforestation, and discharge of industrial waste into the atmosphere, play a pivotal role in modeling climatic change. Fluctuating temperature patterns, changes in precipitation, and also the subsequent loss in biodiversity are few effects of climatic change in agriculture. AIoT tool which can keep track of climatic parameters can assist in tackling hazardous weather conditions [4].

Traditional farming methods often hinder farmers' ability to access current-day practices, trends, and technologies, thereby limiting their capacity to maximize crop yields and sustainability. Relying intensely on manual labor, physical effort, and restricted machinery, traditional farming demands significant capital investment and can sometimes lead to decreased agricultural productivity. Also, the infrastructure in rural areas is poor, including roads, storage facilities, and market links, which can hamper the timely transport and marketing of agricultural output; this leads to post-harvest loss. Continuous cultivation of the land can reduce its fertility due to repetitive farming practices. Burning of fossil fuels releases greenhouse gases, contributing to climatic change and changing weather patterns, which can lead to extreme weather events. Air pollution from burning fuels can directly harm crops, reducing their yield and quality. Acid rain can damage soil quality

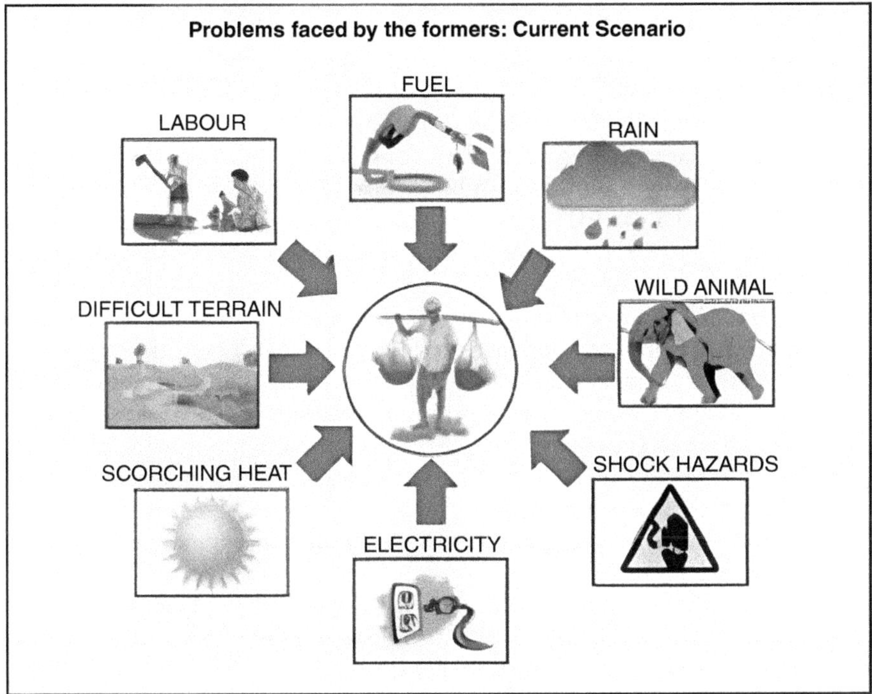

Figure 6.1 Image representing challenges in agriculture.

and affect plant health. The involvement of middlemen in agriculture can lead to reduced profits in agriculture due to higher transaction costs and price manipulation. Lack of transparency between farmer and market can lead to unfair compensation; in addition, the middleman can exploit the knowledge gap. Urbanization can impact agriculture by reducing the use of land for cultivation, leading to a downfall in agricultural productivity. Marginalized traditional practices lead to a loss of traditional knowledge and agricultural techniques [5]. Soil salinity is also a significant challenge in agriculture; high salt concentration will block water uptake, thereby leading to dehydration and ion toxicity in plants. It can also modify soil structure, impend root growth, and affect nutrient absorption, ultimately impacting overall crop health and productivity [6] (Figure 6.1).

6.3 APPLICATIONS OF AIoT

AI has been the tool responsible for the drastic changes happening in the present time and has impacted life in many ways. IoT has been in use for a while now as sensors and other monitoring machines. AI and IoT have been integrated in several fields to make things better with less waste and more benefits. This study focuses on the application of AIoT in agriculture, but there are many different uses of it:

- **Plant Health:** Health is the most precious wealth for a biological entity. Crop management using modern conveniences enables real-time monitoring of crop conditions and alerts on issues detected. The database maintained by AIoT gives consolidating data for farmers and easy access to historical trends. This helps farmers in the early detection and alleviation of crop diseases [7].
- **Smart Storage:** Warehouse plays an active role in post-harvest maintenance. Equipping warehouses with modern AIoT facilities eases crop storage for farmers over extended periods. It enables efficient monitoring and tracking of crops, optimizing storage space, and implementing enhanced security measures through remote monitoring alerts [8].
- **Vehicle Management:** Farming vehicles can be maintained by AIoT sensors by checking them at regular intervals. These sensors alert farmers with blinking indicators whenever maintenance or adjustments are required.
- **Supply Chain Optimization:** AIoT can upgrade supply chain visibility and efficiency. It enables real-time goods tracking and route optimization and reduces cost and delivery time [9].
- **Environmental Monitoring:** AIoT aids in environmental monitoring by collecting and analyzing data on air and water quality as well as weather conditions. This is valuable for early warning systems and conservative efforts.
- **Conserving Wildlife:** AIoT can assist in understanding animal behavior, conserving ecosystems, and preventing poaching. It monitors and protects wildlife with cameras, sensors, and other IoT systems. It supports biodiversity, which co-operates with agriculture in providing natural fertilizers.
- **Energy Optimization:** Energy consumption can be optimized using AIoT by controlling heating, cooling, and lighting systems intelligently. It will be very beneficial for homes, for industries, and at every place where energy consumption is higher. Additionally, it facilitates efficient energy employment in agricultural activities like water pumping.
- **Smart Waste Management:** Waste management is a major concern for preserving Mother Nature, and it has a significant impact on human health as well. The traditional way of garbage collection poses a threat to the environment. AIoT can deal with this concern by providing smart bins and suggesting eco-friendly ways of waste management. Handling waste smartly can reduce pollution and sudden climatic changes, which act as a helping hand for agricultural practices [10] (Figure 6.2).

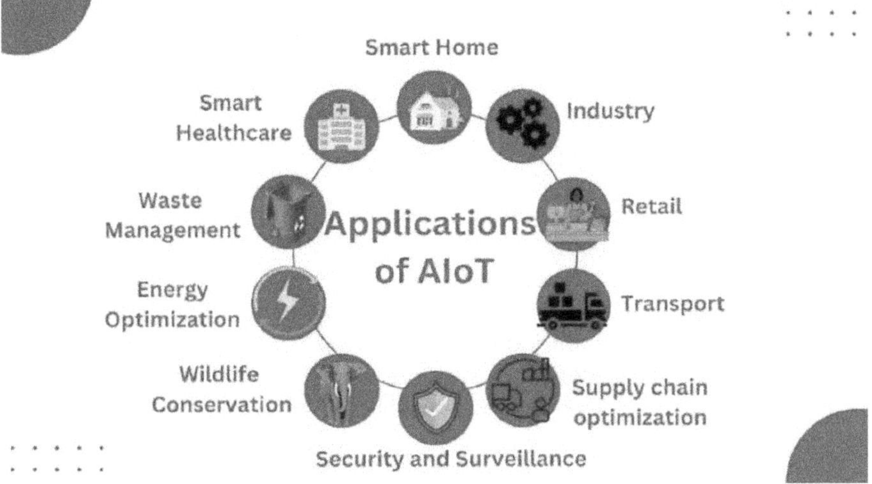

Figure 6.2 Image showing general applications of AIoT in real life.

6.4 WHY AIoT?

Several modern mechanisms such as mechanization, biotechnology, remote sensing, and satellite imagery can be used to modernize agriculture, but AIoT has its own benefits over them. AIoT can recast agriculture in numerous ways, making it more productive, sustainable, and efficient. Mechanization is replacing manual labor with machines like tractors and harvesters. Putting AIoT sensors in tractors will help check the condition of the soil, optimizing their path and reducing fuel consumption. The sensors would also help in sensing the state of the machine's health. Throwing light on the use of biotechnology in agriculture, the traditional modernization technique is to develop genetically modified crops. These crops are resistant to pests and fast in adapting to the environment. Combining AIoT with biotechnology can provide real-time data on crop health and growth, maximizing the potential of these genetically modified seeds. Bringing the method of remote sensing and satellite imagery into consideration, it is basically monitoring crop health and land use. AIoT sensors can give more real-time on-site data that will give more detailed, localized information on crop conditions and allow timely interventions.

6.5 IMPLEMENTING AIoT IN AGRICULTURE

For transforming the traditional methods of agriculture to smart techniques, the correct approach to implementing AIoT is a must. Data collection, sensing present situations, and prediction based on the data collected are the strengths of AIoT. It becomes necessary to use AIoT in such a way that it is beneficial in all aspects.

- **Precision Farming**: Precision farming is managing agricultural practices in a way to get increased yield of crops more efficiently. It involves specific sensing of the farming site/geographical region and sampling. It allows farmers to treat a field as a heterogeneous entity. Inevitably it integrates a full-size quantity of computing and electronics, and better stages of control require an extra state-of-the-art systems approach [11]. Consequently, precision farming is not merely a set of solutions at the local level. It also requires the monitoring and evaluation of agricultural operations both locally and on individual farms, along with a comprehensive understanding of the strategies needed to utilize inputs effectively to achieve specific goals [12].

  AIoT being merged with precision farming will help farmers monitor the growing land and the best suitable conditions for growing a crop. It will help measure the consumption of water and reduce the use of chemicals (help reduce the usage of pesticides and use more organic manure).
  - **Case Study**: In Salinas Valley, California, which is a leading lettuce-growing region, precision farming techniques using AIoT technologies were implemented successfully. Monitoring soil moisture levels, nutrient content, and weather patterns by the deployment of sensors assisted farmers in modifying their irrigation schedules and fertilizer applications. This approach achieved targets like improved crop yields and quality, in addition to significant water savings and reduced environmental impact [7] (Figure 6.3).
- **Data-Driven Agriculture**: Data-driven agriculture supports precision farming by making thoughtful use of the data given by technologies like AIoT. Using the data provided by the sensors, AI can make predictions for supplementing crop yield, reducing crop wastage, and limiting the use of pesticides and chemicals [13].
  - **Case Study**: The data-driven agriculture scheme was introduced by the Columbian Ministry of Agriculture. It used machine learning algorithms to compile and interpret data from various sources. It presented knowledge to farmers through local advisory services. This knowledge enabled farmers to manage agricultural practices effectively for increased and consistent yield [14].

6.5.1 Predictive machinery maintenance

Machinery used in agriculture like tractors, water pumps, and water suppliers for irrigation plays a crucial role in good crop yield. AIoT can help in sensing the health of the machines and warn before machine failure occurs. It can give suggestions for maintaining the machines and suggest ways to

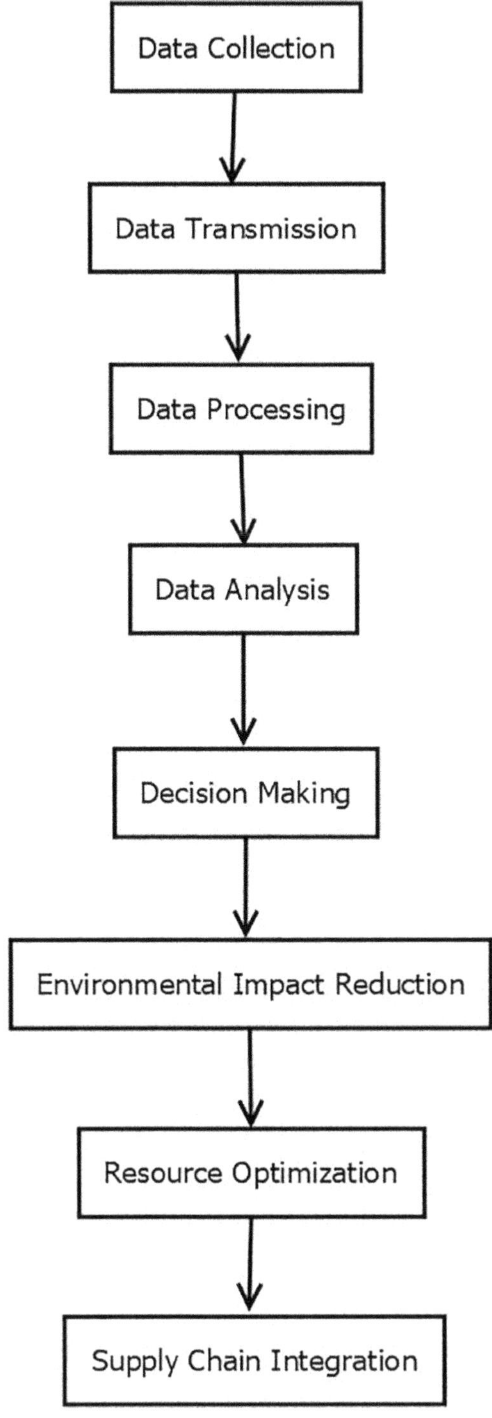

Figure 6.3 Flowchart showing the process of precision farming.

reduce electricity and fuel consumption. Using machines integrated with AIoT like drones can be useful in spraying pesticides, irrigating, and monitoring crops.

- **Crop Monitoring and Yield Productions:** Crop yield may vary from time to time depending on the type of climate, condition of the soil, and other surrounding situations. Technologies like AIoT can help in predicting the climate and give suggestions to protect the crop if there is going to be a change in the climate in the upcoming days. IoT sensors installed in the agricultural land will collect real-time data on temperature, soil moisture, pH, humidity, nitrogen, phosphorous, and other factors and maintain a database of this data. AI algorithms will analyze and identify patterns in the data, predict future trends, and take necessary action against them. These sensors can predict heavy rainfall or heat waves and provide ways to deal with them beforehand. Remote sensing can help monitor crops and warn if any signs of disease are found in them. Tools for frequent soil testing can help monitor the condition of soil and predict ways to preserve or increase the fertility of soil [15]. Smart remote sensors will help count fruitlets, flowers, or fruits at various phases of growth. It will save a lot of time and money compared to physical counting done by manual labor. Estimating the crop yield will help decide whether all the conditions are suitable for the crop or what measures can be taken to increase the yield [16].
- **Supply Chain Optimization:** Producing good quality crops is not enough until the customer receives them in good quality. Supply chain refers to transporting products from supplier to customer. With agriculture becoming smart, the supply chain also needs to be optimized to provide fresh and good products to the customers. If the fruits or vegetables are spoiled during transportation, it is a loss for the farmers and vendors. For optimizing the supply chain, storing products in a manner that they can bear the transportation period and stay as fresh as possible is very important. Smart tools can help store the products in a manner that they can stay fresh for a long time. AIoT can again come into play and can predict for how long the product is going to stay fresh at a particular phase of its growth to decide if it can be transported to what distance before being spoiled [17].
  - **Case Study:** Thailand optimized its rice supply chain to overcome challenges and to make the management of staple food more efficient. The challenges it faced were inefficient transportation, storage, and processing. Using structured transportation routes, it reduced delivery costs. It adapted inventory control systems and updated equipment's enhanced storage capacity and product quality. This brisk response to market requirements improved customer satisfaction [18].

6.5.2 Environmental impact

Observing the condition of the soil and regular monitoring of plants for pests and diseases would be useful to reduce the use of pesticides and other agrichemicals. Less usage of chemicals will be beneficial for the environment and the health of living beings.

- **Risk Management:** Early warning systems can be provided by AIoT that will predict extreme weather events, diseases, or any kind of risks for crops and soil. This will be helpful for the farmers to take quick action before any harm occurs to the crop. These systems will also help provide information about the health of machines before failure occurs.
- **Long-Term Sustainability:** AIoT supports sustainable agriculture so it will complement the farmers in adopting practices that will help preserve natural resources and soil health. It will help store the crops in such a way that they are fresh for a longer time.
- **Labor Efficiency:** Some farmers are not able to afford manual labor because of high costs or deficiency of labor in that area. Automation and robotics in agriculture by implementing AIoT will help reduce manual labor and will be a one-time investment for a number of years. It will be cost effective for the farmers.
- **Soil Health Monitoring:** Intensive tilling, monocropping, and extensive use of chemicals like fertilizers and pesticides will deplete soil nutrients and lead to soil erosion, loss of organic matter, and reduction of water-holding capacity in the soil. Using AIoT devices, soil health can be monitored. This data will help farmers make informed decisions to prevent soil degradation and maintain nutrient balance [15].
- **Optimized Resource Utilization:** In traditional agriculture, resources like water, nutrients, and other pesticides are frequently overused. With the integration of AIoT in agriculture, data-driven recommendations can be leveraged to optimize the overuse of these vital resources and promote sustainable agricultural practices [19].
- **Biodiversity Conservation:** In traditional agriculture, the common practice of monoculture and habitat destruction will present a significant risk to biodiversity. AI can play a crucial role in designing sustainable crop rotation systems and mitigating the adverse effects of ecosystems. IoT devices can monitor and safeguard wildlife, ensuring their protection. Furthermore, AI-enabled cameras can actively detect wildlife movements. By implementing these technologies, a harmonious balance between agricultural productivity and the preservation of the ecosystem can be encouraged, thus promoting a more sustainable approach [20].

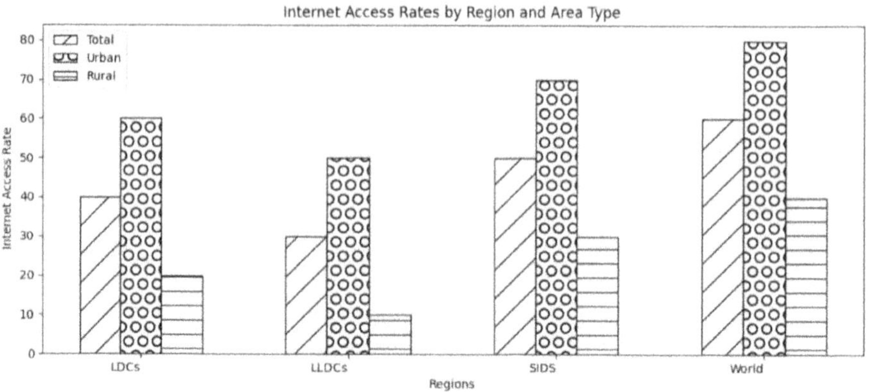

Figure 6.4 The Internet access rate in rural and urban areas.

6.6 CHALLENGES IMPLEMENTING AIoT

Making agriculture smart is a revolutionary idea, and any revolution brings a list of challenges with it. Some of the challenging hurdles of implementing AIoT in agriculture are listed below:

- **Connectivity:** Most of the farms and agricultural lands are located in rural areas where connectivity is still an issue. Though development is happening at a great speed, frequent power cuts and less connectivity still exist in small cities and villages. AIoT cannot be implemented until proper connectivity is attained. The speed of the Internet is also a point of consideration to get the predictions and results in an ample amount of time. Especially for weather forecasting if the Internet is not available, then the warning systems would be inappropriate, and no prior arrangements can be made to face the severe weather conditions. This problem can be solved by deploying mesh networks or satellites alongside traditional infrastructure of the Internet, ensuring connectivity in remote areas (Figure 6.4).
- **Energy Efficiency:** The whole world is trying to increase electricity generation in all ways possible, but rural areas face a deficiency of energy to date. The AIoT devices will need reliable access to power for proper usage. Without energy, the devices will not be able to charge and generate results. Promoting the use of renewable energy resources such as solar panels in farms will be an aid to this problem.
- **Data Integration:** A lot of data will be produced by the AIoT devices. This data needs to be integrated properly for proper usage. The sensors that test soil conditions and their results should be sent to the AI device, which will suggest for preservation, and the weather forecast data should be sent to warning systems to give proper warnings.

Agriculture involves a wide range of data like soil data, crop information, and machinery data. This data is not easy to manage and requires robust pipelines and systems. Proper data integration demands the development of robust data management platforms for seamlessly integrating data from various sources and delivering actionable insights.

- **Security and Data Privacy**: The agricultural data collected will include sensitive information about crops and farm operations. This raises concerns about security and data privacy. Protection of data from theft or cyber-attacks becomes crucial. To ensure the safety of data, implementing encryption protocols and regular security audits is advisable.
- **Interoperability**: Implementing AIoT in agriculture will involve a mix of different sensors and devices from different manufacturers. It will be a significant challenge to ensure that all the systems work together properly. Industrial standards and protocols can be established to guarantee seamless interoperability between the devices.
- **Scalability**: The AIoT solutions should be highly scalable because agricultural operations may vary greatly in size. The solutions should be executable by both small family farms and large commercial operations. Designing modular AIoT solutions that are capable of scaling according to the size of operation with flexible deployment options can be helpful.
- **Cost**: The biggest challenge in implementing AIoT in agriculture will be cost. AIoT devices will definitely cost more than manual labor and physical practices for once. But once money is invested in such devices, they will provide service for years and end up saving money and time. Offering subsidies to farmers for adopting AIoT devices, coupled with subscription models, can be beneficial.
- **Training**: The devices will need to be handled in a particular way, and the person operating it should understand their functioning. The farmers and agricultural workers need to be trained, and this would be a time-consuming and costly process. The use of these devices would seem complex for some elderly farmers and workers. However, the elder generation of farmers would face difficulties and may not like these devices. The new generation or the next generation who will follow their ancestral occupation of farming would be curious and eager to learn new techniques. Customized training programs that emphasize user-friendly interfaces and hands-on learning can assist farmers and agricultural workers.
- **Regulation**: Government regulations are followed in agricultural practices, which may not keep pace with AIoT developments. Compiling these regulations while implementing AIoT can be very challenging. Collaborating with regulatory bodies to adapt frameworks that accommodate AIoT technologies ensuring compliance with existing regulations will support overcoming this challenge.
- **Reliability and Maintenance**: Agriculture is one of the most important aspects of economy and health, and it cannot have any kind of

downtime. Systems must be reliable enough to reduce the chances of faults. Any failure in agricultural operations would be very expensive for the farmers and for the economy too. Maintenance of the sensors and other systems is very important, which can be a challenge for remote areas. Utilizing AIoT sensors and analytics to implement predictive maintenance prevents downtime of devices.

- **Sharing and Owning Data:** Ownership of data would be a questioning aspect like the whole data should be owned by the farmers or other stakeholders as well. Sharing of data between farmers and other stakeholders like technology providers and researchers would be a contentious issue. It is important to establish clear regulations around data sharing agreements and ownership rights, which should include fair compensation and privacy protection for all stakeholders.
- **Design and Disposal:** Designing the systems in an agricultural-friendly way is very important. The systems should not harm the plants or crops while testing; they should not produce a lot of electronic waste, and they prompt if they are giving an approximate result. The disposal of the sensors and system should be done carefully so that they don't cause any harm to the environment. Promotion of recycling and proper disposal of AIoT hardware can be supportive.
- **Resistance to Adoption:** Farmers and other agricultural workers who have been practicing traditional methods of agriculture may resist adopting new technologies. People will not be ready to take the risk of adopting new techniques in their occupations. Spreading awareness becomes crucial for making agriculture smart. Conducting regular public demonstrations on AIoT technologies will foster confidence in farmers to adopt these technologies.

6.7 AWARENESS AMONG FARMERS

Raising awareness among farmers is as essential as the development of agricultural technologies. Several communication channels can effectively reach them, which includes advertising on television, radio [21], social media [22], newspaper articles, and many more. An impactful approach includes establishing a committee of four to six dedicated individuals, who can periodically visit farmers in person. Through these visits, they can provide personalized explanations and training sessions on the utilization of modern agricultural technologies. This approach not only facilitates direct communication but also allows for a more comprehensive understanding of specific needs and challenges and removes ambiguities for farmers. By imparting knowledge and practical training, this committee can bridge the gap between technology and its practical application in the agricultural landscape. This method also promotes a sense of community and collaboration, encouraging farmers to adopt these advancements and effectively integrate them into farming practices, thus ensuring sustainable agricultural development and enhanced productivity.

The "Online Farming Program" in collaboration with information technology is an example of one of the awareness strategies used to promote the adoption of AIoT in agriculture. This program utilizes digital tools and resources to present practical knowledge. This initiative focuses on bringing a tech-savvy agricultural community, empowering farmers to make reliable decisions to maximize their yields. Farmers have gained access to technologies tailored to their needs by real-world adoption of this strategy. The initiative comprises technologies such as soil sensors and crop monitoring, which made the farmers realize the potential of AIoT.

Efforts to raise awareness about AIoT in agriculture include a collection of strategies; we can include research institutes to conduct surveys and studies and then advertise the data, highlighting the advantages of AIoT. We can arrange on-site presentations showcasing the practical applications of AIoT devices and their positive effects on farming activities. Educating farmers can also include translating educational materials into local languages, enabling better understanding among farming associations. Furthermore, there is a need to advocate for government subsidies [18] to support farmers investing in modern technologies. Encouraging farmers to establish a network for them facilitates the exchange of their valuable thoughts, insights, and experience related to technology implementation. Integrating AIoT farming approach into educational institutions familiarizes the next-generation farmers with the benefits of these technologies. Encouraging farmers' active participation in seminars to share their practical experience in implementing AIoT can motivate their fellow farmers. Showcasing how AIoT can effectively mitigate the obstacles faced by traditional farming experience. Moreover, highlighting the long-term positive impact of AIoT in agriculture will encourage more people to accept and use these new technologies. Hosting seminars and conferences to showcase the latest advancements and trends in AIoT for agriculture can significantly contribute to enhancing understanding and spreading awareness among stakeholders. In conclusion, raising awareness is the foundational step toward improved implementation of these technologies.

6.8 CONCLUSION

Practicing agriculture with less wastage of crops and more productivity has become crucial worldwide due to the increasing demand for food. Technologies like AI and IoT have complimented many occupations, and the combination of these two technologies, AIoT, is proving to be a boon for the modern era. AIoT is best suited to overcome the challenges in agriculture and supplement farmers in fulfilling worldwide demands for food. Farmers practicing the traditional method of farming will definitely think before bringing any new change to their occupation. It will not be easy for everyone in the field of agriculture to accept changes in a short span of time. Implementing AIoT in agriculture will not just promote efficient farming but will also provide ways for the long-term sustainability of agricultural products. Once this change is accepted by farmers, it brushes up the productivity on their agricultural land, and they will end up producing good quality crops with very little wastage. Improved agricultural practices are not only restricted to the benefits of farmers but valuable for countries having agriculture as one of the main occupations. Implementing AIoT in agriculture will boost the economy of such countries and also benefit the health of people.

To grasp the potential of AIoT in agricultural practices to the fullest, combined efforts are needed to address present challenges and propel future advancements. Refining AIoT algorithms for enhanced predictive capability, optimized sensing, and measures to safeguard sensitive agricultural data should be focused on in future research. Collaborations of technology providers, farming communities, government agencies, and academic institutions can facilitate technology adoption. The transformative power of AIoT will raise food productivity that ensures food security for future generations. It's time to unite and take action to move toward a smart and more reliable agricultural system.

ACKNOWLEDGMENTS

We extend our deepest gratitude to everyone who contributed to this chapter. Special thanks to MS Ramaiah University of Applied Sciences for its invaluable resources and support. We appreciate the insightful feedback from reviewers and the suggestions from our colleagues and friends. We also thank the publishers for their professionalism and dedication in bringing this work to a wider audience. Lastly, we recognize all uncredited individuals, whose expertise, collaboration, and encouragement have been essential in completing this work.

REFERENCES

1. Rothstein SJ. Returning to our roots: making plant biology research relevant to future challenges in agriculture. *The Plant Cell.* 2007;19(9):2695–9.
2. Van Dijk M, Morley T, Rau ML, Saghai Y. A meta-analysis of projected global food demand and population at risk of hunger for the period 2010–2050. *Nature Food.* 2021;2(7):494–501.
3. Nozari H, Szmelter-Jarosz A, Ghahremani-Nahr J. Analysis of the challenges of artificial intelligence of things (AIoT) for the smart supply chain (case study: FMCG industries). *Sensors.* 2022;22(8):2931.

4. Devendra C. *Climate Change Threats and Effects: Challenges for Agriculture and Food Security*. Kuala Lumpur: Academy of Sciences Malaysia; 2012.

5. Oenema O, Velthof G, Kuikman P. Technical and policy aspects of strategies to decrease greenhouse gas emissions from agriculture. *Nutrient Cycling in Agroecosystems*. 2001;60:301–15.

6. Mukhopadhyay R, Sarkar B, Jat HS, Sharma PC, Bolan NS. Soil salinity under climate change: Challenges for sustainable agriculture and food security. *Journal of Environmental Management*. 2021;280:111736.

7. Smukler SM, Jackson LE, Murphree L, Yokota R, Koike ST, Smith RF. Transition to large-scale organic vegetable production in the Salinas Valley, California. *Agriculture, Ecosystems & Environment*. 2008;126(3–4):168–88.

8. Fichtinger J, Ries JM, Grosse EH, Baker P. Assessing the environmental impact of integrated inventory and warehouse management. *International Journal of Production Economics*. 2015;170:717–29.

9. Wu CK, He Y, Tsang KF, Mozar S. The IDex case study on the safety measures of AIoT-based railway infrastructures. In *2020 IEEE International Symposium on Product Compliance Engineering-Asia (ISPCE-CN) 2020* (pp. 1–4). Chongqing, China: IEEE.

10. Sung WT, Devi IV, Hsiao SJ, Fadillah FN. Smart garbage bin based on AIoT. *Intelligent Automation & Soft Computing*, 2022;32(3):1387–401.

11. Finger R, Swinton SM, El Benni N, Walter A. Precision farming at the nexus of agricultural production and the environment. *Annual Review of Resource Economics*. 2019;11(1):313–35.

12. Blackmore S. Precision farming: an introduction. *Outlook on Agriculture*. 1994;23(4):275–80.

13. Gupta S, Kumar D, Mishra S. Big data analytics in agriculture: Harnessing IoT-generated insights. In *Agriculture 4.0* (pp. 185–201). CRC Press.

14. Jiménez D, Delerce S, Dorado H, Cock J, Muñoz LA, Agamez A, Jarvis A. A scalable scheme to implement data-driven agriculture for small-scale farmers. *Global Food Security*. 2019;23:256–66.

15. Joshi-Bag S, Vyas A. AIoTST-CR: AIoT based soil testing and crop recommendation to improve yield. *Indonesian Journal of Electrical Engineering and Informatics (IJEEI)*. 2023;11(3):750–63.

16. Darwin B, Dharmaraj P, Prince S, Popescu DE, Hemanth DJ. Recognition of bloom/yield in crop images using deep learning models for smart agriculture: A review. *Agronomy*. 2021;11(4):646.

17. Khandelwal C, Singhal M, Gaurav G, Dangayach GS, Meena ML. Agriculture supply chain management: A review (2010–2020). *Materials Today: Proceedings*. 2021;47:3144–53.

18. Wilasinee S, Imran A, Athapol N. Optimization of rice supply chain in Thailand: a case study of two rice mills. *Sustainability in Food and Water: An Asian Perspective*. 2010:263–80.

19. Davijani MH, Banihabib ME, Anvar AN, Hashemi SR. Optimization model for the allocation of water resources based on the maximization of employment in the agriculture and industry sectors. *Journal of Hydrology*. 2016;533:430–8.

20. Reed BM, Sarasan V, Kane M, Bunn E, Pence VC. Biodiversity conservation and conservation biotechnology tools. *Vitro Cellular & Developmental Biology-Plant*. 2011;47:1–4.

21. Nwagwu WE. Farmers' awareness and use of information and communications technologies in the livestock innovation chain in Ibadan City, Nigeria. *Mousaion*. 2015;33(4):106–30.
22. Maulani G, Rahardja U, Hardini M, I'zzaty RD, Aini Q, Santoso NP. Educating farmers using participatory rural appraisal construct. In *2020 Fifth International Conference on Informatics and Computing (ICIC)*, 2020 Nov 3 (pp. 1–8). Indonesia: IEEE.

## Chapter 7

Enhancing medical practice with AI innovations in diagnostics and treatment

K. Sumathi, P. Pandiaraja, K. Muthumanickam, and P. Rajesh Kanna

7.1 INTRODUCTION

Artificial intelligence (AI) refers to the simulation of human intelligence processes by machines, especially computer systems [1,2]. These processes include learning (the acquisition of information and rules for using that information), reasoning (using rules to reach conclusions or solve problems), and self-correction [3].

7.1.1 Types of AI

AI is a multidisciplinary field that encompasses various subfields, such as machine learning, natural language processing (NLP), computer vision, and robotics. The goal of AI is to create machines that can perform tasks that typically require human intelligence, such as understanding natural language, recognizing patterns in data, making decisions, and even exhibiting some level of creativity (Figure 7.1).

- **Narrow or Weak AI:** This type of AI is designed and trained for a specific task, such as voice assistants like Siri or recommendation systems on streaming platforms. It excels in its specific domain but lacks general intelligence.
- **General or Strong AI:** This refers to AI that possesses human-like cognitive abilities and can perform any intellectual task that a human

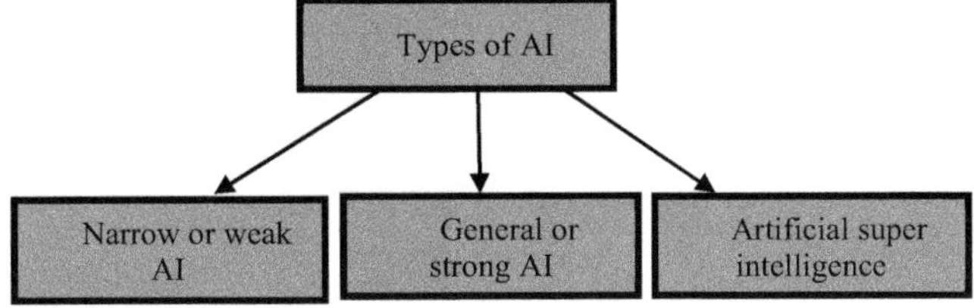

Figure 7.1 Types of AI.

DOI: 10.1201/9781032618173-9

being can. It can realize, be trained, and apply knowledge across a broad choice of responsibilities, potentially surpassing human capabilities.

- **Artificial Super Intelligence:** This hypothetical level of AI surpasses human intelligence in every aspect and has the potential to perform tasks far beyond human capabilities. It is currently a topic of philosophical and ethical debate.

7.1.2 Implications of AI across industries

AI has far-reaching implications across various industries, including healthcare, finance, entertainment, transportation, and more. However, the rapid advancement of AI also raises important ethical, societal, and regulatory considerations, such as job displacement, bias in algorithms, privacy concerns, and the impact on human decision-making (Figure 7.2).

7.1.3 Application of AI

AI has a broad variety of applications across a variety of industries and domains. The large volume of information is considered for analyzing and predicting some content for automation as well as, learned from experience, creates considerable advancement in a lot of fields. The most important applications of AI are listed as follows:

- **Healthcare:**
  - **Medical Imaging Analysis:** AI supports predicting diseases by analyzing medical images like X-rays, MRIs, and CT scans and finding abnormalities [4–7]. *Drug Discovery:* AI algorithms can guess possible remedy candidates and their exchanges, accelerating the drug invention practice. *Personalized Treatment:* AI helps adapt treatment plans based on particular patient data and their medical histories [8,9]. *Remote Monitoring:* AI-enabled devices are able to observe patients' imperative signs and are ready to act as healthcare providers in real time [10,11].

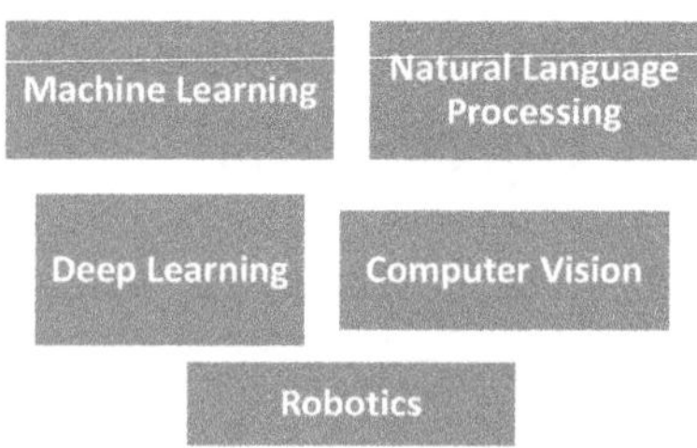

Figure 7.2 AI techniques.

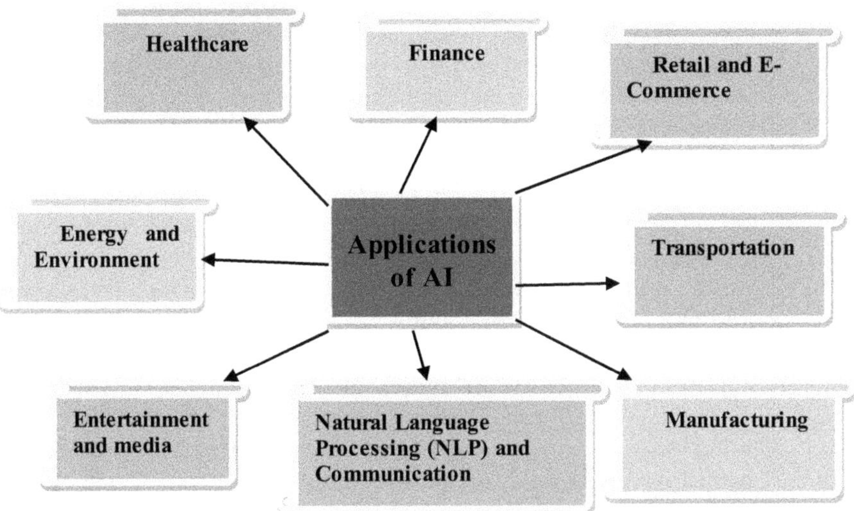

Figure 7.3 Applications of AI.

- **Finance:** *Algorithmic Trading*: AI algorithms examine economic-related information to formulate quick trading decisions. *Fraud Detection*: AI detects patterns pinpointing fake activities in economic dealings. *Credit Scoring*: AI assesses credit risk by analyzing consumer-related information and their financial history. *Robo-Advisors*: AI-powered systems make available computerized savings guidance to the consumer based on their goals and risk kindness (Figure 7.3).
- **Retail and E-Commerce:** *Recommendation Systems*: AI suggests certain products to consumers based on their predilection and procure records. *Inventory Management*: AI optimizes inventory levels and supply chain operations. *Chatbots*: AI-driven chatbots give client support and help by investigation. *Visual Search*: AI enables users to seek goods using images rather than content.
- **Transportation:** *Autonomous Vehicles*: AI-powered automated vehicles and trucks facilitate protected and more proficient transportation. *Traffic Management*: AI optimizes traffic flow and diminishes overcrowding in metropolitan areas. *Predictive Maintenance*: AI evaluates statistics from automobiles and machines to guess preservation desires in advance and avoid breakdown.
- **Manufacturing:** *Quality Control*: AI systems perceive defects and abnormalities in industrialized development. *Predictive Maintenance*: AI foresees utensil malfunctions, decreasing downtime and increasing efficiency. *Process Optimization*: AI optimizes production processes for improved efficiency and resource utilization.

- **NLP and Communication:** *Language Translation*: AI-powered tools decipher verbal and non-verbal communication in real time. *Chatbots and Virtual Assistants*: AI chatbots handle client inquiries and execute errands through verbal communications. *Sentiment Analysis*: AI analyzes social media and customer feedback to gauge public sentiment.
- **Entertainment and Media:** *Content Recommendation*: AI suggests movies, music, and articles based on user preferences. *Content Creation*: AI generates written content, artwork, and music. *Video and Image Analysis*: AI identifies objects, scenes, and faces in images and videos.
- **Energy and Environment:** *Energy Management*: AI optimizes energy consumption in buildings and industries. *Renewable Energy*: AI forecasts energy production from solar and wind sources for efficient grid management. *Environmental Monitoring*: AI analyzes data to monitor air and water quality and predict natural disasters [12].

7.2 RELATED WORK

AI in healthcare is a rapidly evolving field with numerous research papers and related work. Here are some important topics and notable papers in AI healthcare.

- "Deep Residual Learning for Image Recognition" introduced ResNets, a revolutionary neural network architecture that addresses the vanishing gradient problem, allowing for the training of extremely deep networks [13]. This innovation has had a significant impact on various domains, including medical image analysis, where it has led to improved accuracy and efficiency in diagnosing and analyzing medical images. ResNets continue to be a cornerstone in the field of deep learning and computer vision, driving advancements in image recognition and analysis.
- "CheXNet: Radiologist-Level Pneumonia Detection on Chest X-Rays with Deep Learning" demonstrates the power of deep learning in diagnosing pneumonia from chest X-ray images, offering the potential to augment and expedite the work of radiologists and improve the accuracy and accessibility of healthcare diagnostics [14].
- "EfficientNet: Rethinking Model Scaling for Convolutional Neural Networks" presents a novel approach to scaling neural networks, offering a family of models that balance computational efficiency and accuracy [15]. This innovation has significant implications for medical image analysis, enabling the deployment of efficient yet accurate models in healthcare and beyond.

- "DeepChem: A Genome-scale Chemoinformatics Library" introduces a transformative platform that leverages deep learning and chemoinformatics to expedite drug discovery and advance precision medicine [16]. With its open-source nature, flexibility, and collaborative approach, DeepChem has the potential to reshape the landscape of pharmaceutical research and ultimately improve healthcare by facilitating the development of more effective and personalized treatments.

AI-assisted surgery has continued to evolve and gain acceptance in various surgical specialties, including orthopedics [17]. Robotic systems have become increasingly sophisticated, and surgeons have further embraced these technologies to provide safer and more effective patient care. The study by Mann et al. paved the way for ongoing advancements in AI-assisted surgery and its application in other surgical disciplines.

7.3 APPLICATION OF AI IN HEALTHCARE

AI is making major contributions to the healthcare sector, modernizing how healthcare experts make a diagnosis, treat, and supervise patient care. The forthcoming section describes some key applications of AI in the healthcare sector:

1. Medical Imaging Analysis:
   - **Diagnosis and Detection**: AI algorithms examine medical images (such as X-rays, MRIs, and CT scans) to detect irregularity, tumors, and other circumstances with high precision.
   - **Early Disease Detection**: AI can recognize premature symbols of sickness like tumors, facilitating judicious intrusion in advance.
   - **Radiology Assistance**: AI assists radiologists by highlighting potential areas of concern, reducing human error.
2. Drug Discovery and Development:
   - **Virtual Screening**: AI models predict potential drug candidates by analyzing molecular structures and their interactions.
   - **Optimized Drug Design**: AI helps design molecules with desired properties and minimal side effects.
   - **Clinical Trial Design**: AI assists in designing efficient and targeted clinical trials.
3. Personalized Treatment and Care:
   - **Precision Medicine**: AI analyzes patient data to tailor treatments based on individual genetics, medical history, and responses to therapy.
   - **Treatment Recommendations**: AI suggests optimal treatment plans by considering the latest medical research and patient data.
4. Healthcare Operations and Management:
   - **Predictive Analytics**: AI predicts patient admission rates, resource needs, and disease outbreaks, aiding hospital planning.

- **Supply Chain Management:** AI optimizes inventory levels of medical supplies and equipment.
- **Fraud Detection:** AI identifies fraudulent insurance claims and billing irregularities.
- The application of AI in healthcare is continuously evolving, and it holds the potential to improve patient outcomes, enhance clinical decision-making, reduce costs, and ultimately transform the way healthcare is delivered. However, it's important to address ethical and regulatory considerations, such as patient privacy, transparency in algorithms, and proper validation of AI-driven medical solutions.

7.3.1 AI support in medical imaging analysis

AI has significantly advanced medical imaging analysis by providing tools and techniques that enhance the accuracy, efficiency, and speed of diagnosis and interpretation. Here's how AI is supporting medical imaging analysis.

7.3.1.1 Image recognition and classification

AI algorithms can automatically identify and classify structures and anomalies in medical images. For example, they can identify different types of tumors, fractures, or lesions in X-rays or MRIs.

7.3.1.2 Segmentation and anatomical mapping

The medical images are segregated as organs, tissues, or structures by an AI tool, which helps to quantify deviations in those images and determine their proportions.

7.3.1.3 Detection of anomalies and abnormalities

AI algorithms are able to flag perspective anomalies in the medical images, which help radiologists spotlight significant conclusions.

7.3.1.4 Early disease detection

AI is capable of perceiving slight changes in medical images that may point out early phases of diseases like cancer, earlier than they are noticeable to the individual.

7.3.1.5 Pattern recognition and diagnosis

The medical image patterns and features are recognized by AI tools, which enables to analyze diseases, guess outcomes, and give the best treatment preferences.

7.3.1.6 Comparison and progression analysis

AI tools have the ability to compare recent medical images with existing medical images and find out disease sequences or handling efficiency over time.

7.3.1.7 Quantitative analysis

AI facilitates exact dimensions of structures and tissues, which preserve support in scrutinizing surroundings and action comeback.

7.3.1.8 Automated report generation

The preliminary radiology reports are generated by the AT tool by pulling out appropriate information from medical images and mingling it with patient data.

7.3.1.9 Assisting radiologists

AI serves as a "second opinion" tool, serving radiologists by diminishing possible anomalies and providing that approach for explanation.

7.3.1.10 Enhanced visualization

AI knows how to improve the eminence of medical images by dropping noise, humanizing disparity, and grinding information.

7.3.1.11 Workflow efficiency

AI speeds up the scrutiny procedure, permitting radiologists to be the center of attention on complex cases and reducing analysis points.

7.3.1.12 Multi-modality integration

To provide a complete view of a patient's circumstance, AI preserves integrating huge information from imaging modalities (such as MRI, CT, and PET scans).

7.3.1.13 Training and education

AI-powered tools know how to train students in the educational field.

7.3.1.14 Research and data analysis

AI supports examining huge datasets of medical images for research purposes, uncovering imminent correlations that might be complicated to discover physically.

7.3.2 Tools and techniques used for medical imaging analysis

The AI and machine learning tools are used to analyze medical images. This topic describes some of the well-known tools and techniques used in medical imaging analysis:

7.3.2.1 Convolutional neural networks

Convolutional neural networks (CNNs) are deep learning architectures exclusively designed for image analysis. The primary task of this tool is to classify the image, detect the objects, and make a segment in medical images.

7.3.2.2 Segmentation algorithms

These algorithms are mainly designed to partition a medical image into different sections or sectors; U-Net and Mask R-CNN techniques are used to analyze unambiguous structures or anomalies in the given medical image.

7.3.2.3 Transfer learning

If labeled medical data have limited features, then transfer learning techniques are used to fine-tune them with the help of pre-trained neural network models of large datasets.

7.3.2.4 Random forests and decision trees

These are some of the machine learning techniques that are used to segment the image, to extract some specific features, and for classification tasks in medical images.

7.3.2.5 Support vector machines

The medical images are taken into account, and support vector machines (SVMs) are used for binary classification tasks such as distinctive findings between normal and abnormal.

7.3.2.6 Generative adversarial networks

Generative adversarial networks can create artificial medical images, increase image quality, and assist in data expansion for training models.

7.3.2.7 Image registration

This technique supports several images of the same patient or structure, facilitating assessment and following variation over time.

7.3.2.8 Texture analysis

Texture analysis methods quantify the samples and surfaces present in images, providing supplementary information for judgment and categorization.

7.3.2.9 Radiomics

This method pulls out a large volume of quantitative features from medical images, which can then be used for disease diagnosis, prognosis, and treatment response prediction.

7.3.2.10 Deep reinforcement learning

Deep reinforcement learning is particularly effective for sequential decision-making tasks in medical imaging. During the scanning process, it helps optimize the sequence of imaging views to achieve the most informative results while minimizing unnecessary exposures. By learning from past decisions and outcomes, this technique enhances the accuracy and efficiency of imaging procedures, paving the way for improved diagnostic workflows and patient care.

7.3.2.11 Graph-based methods

Graph theory can be applied to model relationships between different regions in medical images, aiding in understanding complex structures and their interactions.

7.3.2.12 Atlas-based methods

These methods involve aligning a patient's image to a standard anatomical atlas, enabling better localization and comparison of structures.

7.3.2.13 Image fusion

Image fusion combines information from multiple imaging modalities to provide a more comprehensive view of a patient's condition.

7.3.2.14 Quantitative imaging

Quantitative imaging involves measuring specific features in medical images, such as tumor volume or blood flow, to aid in diagnosis and treatment planning.

7.3.2.15 AI platforms

There are specialized platforms and frameworks, such as tensor flow, PyTorch, and digital imaging and communications in medicine libraries that provide tools for building and deploying AI models for medical imaging analysis.

7.3.3 AI support in drug discovery and development

AI has brought about significant advancements in drug discovery and development, accelerating the process of identifying potential drug candidates, optimizing their properties, and streamlining clinical trials [16,18].

7.3.3.1 Virtual screening and compound design

AI models are used to analyze huge compound libraries to suspect which molecules are expected to interrelate with exact disease targets. The drug-utilized candidates are efficiently identified using this method.

7.3.3.2 Predicting drug–target interactions

AI techniques are used to predict communications between drug molecules and genetic targets, give support to understanding drug mechanisms, and recognize probable side effects.

7.3.3.3 Optimizing drug properties

AI-driven algorithms optimize drug properties, such as solubility, bioavailability, and metabolic stability, to increase the chances of success in clinical trials.

7.3.3.4 Personalized medicine

AI analyzes hereditary and molecular data to guess entity reactions to drugs, permitting custom-made treatment policies.

7.3.3.5 Repurposing existing drugs

AI techniques are used to identify new uplifting uses of presented drugs by scrutinizing their relations with different targets, potentially speeding up the treatment processes.

7.3.3.6 Biomarker discovery

AI helps identify biomarkers associated with diseases, aiding in patient selection for clinical trials and monitoring treatment responses.

7.3.3.7 Clinical trial optimization

AI analyzes patient data to optimize clinical trial design, predict patient recruitment rates, and identify potential risks.

7.3.3.8 Drug toxicity prediction

AI models predict the potential toxic effects of drug candidates, reducing the likelihood of adverse reactions in clinical trials.

7.3.3.9 Data mining and knowledge discovery

AI extracts insights from vast datasets, scientific literature, and databases to inform drug discovery strategies.

7.3.3.10 Natural language processing

NLP techniques help researchers extract relevant information from scientific literature, patents, and clinical trial data, aiding in hypothesis generation and decision-making.

7.3.3.11 Chemical synthesis planning

AI designs efficient routes for the chemical synthesis of drug candidates, saving time and resources.

7.3.3.12 High-throughput screening

AI automates and enhances the analysis of large-scale experiments, enabling the screening of thousands of compounds for potential drug activity.

7.3.3.13 Drug formulation and delivery

AI optimizes drug formulations and delivery systems to improve patient compliance and efficacy.

7.3.3.14 Collaborative platforms

AI-powered platforms facilitate collaboration and data sharing among researchers, leading to more efficient drug discovery efforts.

7.3.3.15 Drug regulatory compliance

AI assists in ensuring that drug candidates meet regulatory standards and guidelines. AI's role in drug discovery and development holds the potential to reduce costs, increase success rates, and accelerate the timeline for bringing new treatments to market.

7.3.4 Tools and techniques used in drug discovery and development

This topic describes a variety of tools and techniques based on AI and machine learning used in drug discovery and development to support various stages of the process.

7.3.4.1 Virtual screening and compound design

- **Molecular Docking Software:** Tools like AutoDock and Vina perform virtual docking simulations to predict how small molecules interact with target proteins.
- **Quantitative Structure–Activity Relationship (QSAR) Models:** These models correlate the chemical structure of compounds with their biological activity, aiding in predicting the activity of new compounds.

7.3.4.2 Predicting drug–target interactions

- **Machine Learning Models:** Algorithms such as random forests, SVMs, and deep learning models predict interactions between drugs and target proteins based on molecular features.
- **Network Analysis:** Graph-based approaches analyze protein interaction networks to predict potential drug targets.

7.3.4.3 Optimizing drug properties

- **Quantum Chemistry Calculations:** These simulations compute molecular properties at a quantum level, aiding in predicting properties like solubility and stability.

7.3.4.4 AI-driven property prediction models

Machine learning models predict drug properties like toxicity, solubility, and bioavailability based on molecular features.

7.3.4.5 Personalized medicine

- **Genetic and Omics Data Analysis:** AI analyzes genetic, genomic, and proteomic data to identify biomarkers and predict individual responses to drugs.
- **Pharmacogenomics Models:** These models correlate genetic variations with drug responses to guide personalized treatment choices.

7.3.4.6 Biomarker discovery

- **Feature Selection Algorithms:** Machine learning algorithms help identify relevant features or biomarkers associated with disease states.

- **Dimensionality Reduction Techniques:** Methods like principal component analysis reduce complex data while preserving relevant information.

7.3.4.7 Clinical trial optimization

- **Patient Recruitment Tools:** AI platforms analyze patient data to identify suitable candidates for clinical trials, optimizing enrollment.
- **Predictive Analytics:** AI predicts patient recruitment rates, trial outcomes, and potential risks based on historical data.

7.3.4.8 Drug toxicity prediction

- **Toxicity Prediction Models:** Machine learning models assess potential toxicity based on chemical structure and biological data.
- **Adverse Event Analysis:** AI analyzes clinical trials and real-world data to identify potential adverse reactions.

7.3.4.9 Data mining and knowledge discovery

- **Text and Data Mining Tools:** AI-driven platforms extract relevant information from scientific literature, patents, and databases for insights and hypothesis generation.

7.3.4.10 Natural language processing

- **NLP Libraries:** Tools like NLTK, SpaCy, and BERT process and analyze textual data from scientific literature and clinical trial reports.

7.3.4.11 Chemical synthesis planning

- **Reaction Prediction Models:** AI predicts chemical reactions and proposes efficient synthesis routes for drug candidates.
- **Retrosynthetic Analysis Tools:** These tools suggest steps to synthesize a target compound from available starting materials.

7.3.4.12 High-throughput screening

- **Laboratory Automation:** Robotic systems and automated assays process large volumes of compounds for high-throughput screening.
- **Image Analysis Algorithms:** AI analyzes images of biological assays to quantify results and identify active compounds.

7.3.4.13 Drug formulation and delivery

- **Pharmacokinetics Models:** AI-driven models predict drug absorption, distribution, metabolism, and excretion properties.

- **Formulation Optimization Software:** Tools optimize drug formulations for improved delivery and bioavailability.

7.3.4.14 Collaborative platforms

- **Data Sharing and Collaboration Tools:** Cloud-based platforms and collaboration software facilitate data sharing, analysis, and communication among research teams.

7.3.4.15 Drug regulatory compliance

- **Regulatory Intelligence Platforms:** AI systems monitor and analyze regulatory guidelines and requirements to ensure compliance during drug development.

These tools and techniques demonstrate the diversity of AI-driven approaches in drug discovery and development, supporting researchers in identifying promising drug candidates, optimizing their properties, and navigating the complex process of bringing new treatments to market.

7.3.5 AI support in personalized treatment and care

AI has made significant contributions to personalized treatment and care in various fields, particularly in healthcare. Here are some ways AI supports personalized treatment and care:

- **Medical Diagnostics and Imaging:** AI algorithms can analyze medical images such as X-rays, MRIs, and CT scans to assist in early and accurate diagnosis of diseases. They can identify subtle samples and abnormalities that might not be recognized by human eyes, enabling custom-made and targeted treatment plans.
- **Genomic Analysis:** Usually some diseases are spread based on the genetic nature of human beings; AI tools can analyze those genetic data and their deviations. This information may be useful to the doctors who treat that kind of patients.
- **Drug Discovery:** AI tools are used to speed up the drug discovery process by imitating molecular interactions, guessing drug candidates, and discovering probable side effects. This information may be utilized by doctors for developing personalized prescriptions and remedies.
- **Predictive Analytics:** AI algorithms can examine a patient's medical history, present health conditions, and determine disease exacerbations or complications. This enables positive interferences and adapts caring strategies for the patients.
- **Treatment Recommendations:** AI techniques are used to examine a huge volume of medical history as well as genetic diseases of the patients and recommend the most suitable treatment plans to them.

- **Remote Monitoring and Telehealth:** AI-enabled devices and wearable techniques are used to continuously monitor the patient's health conditions. This information may be useful to both patients and doctors to enable early intervention and alternative treatment plans.
- **Personalized Behavioral Interventions:** AI can analyze patients' behaviors and priorities to build up adapted behavior change interventions, such as encouraging healthy habits, organizing chronic conditions, and humanizing mental health.
- **Surgical Planning and Robotics:** Robots and AI-enabled surgical tools are used in the medical field to enhance the correctness and accuracy of surgical procedures. This scenario may be used by the surgeons at the time of operations.
- **Electronic Health Records (EHRs):** AI can explore EHRs to categorize trends, correspondences, and prospective threats that might guide adapted treatment assessments.
- **Clinical Trials and Research:** AI can assist in classifying proper candidates for clinical trials, optimize trial designs, and examine testing data more efficiently, leading to more targeted and personalized treatments.

7.3.6 Tools and techniques used in personalized treatment and care

Personalized treatment and care in healthcare involve the use of various tools and techniques, including advanced technologies and data-driven approaches.

7.3.6.1 Genomic sequencing and analysis

- **Next-Generation Sequencing (NGS):** NGS technologies facilitate the speedy and commercial sequencing of an individual's complete genome structure or explicit genes, recognizing hereditary variations that may possibly vigor infection hazard and action reply.
- **Bioinformatics:** A number of bioinformatics tools can be used to analyze hereditary information to discover genetic mutations, deviations, and biomarkers related to diseases and treatment responses.

7.3.6.2 Medical imaging and diagnostics

- **Computer-Aided Diagnosis (CAD):** CAD systems use AI algorithms to evaluate medical images (X-rays, MRIs, and CT scans) and discover anomalies, supporting radiologists and clinicians in exact judgment.
- **Imaging Biomarkers:** The specific features in medical images can be identified with the help of quantitative imaging techniques, which help to support disease progression or treatment procedures.

7.3.6.3 Predictive analytics and machine learning

- **Predictive Models:** The patient's data can be analyzed using machine learning algorithms to predict disease risk, treatment outcomes, and probable hurdles.
- **Clinical Decision Support Systems:** AI-powered tools propose treatment recommendations and support clinicians in building informed judgments based on patient information and medical history.

7.3.6.4 Remote monitoring and wearable technologies

- **Wearable Devices:** Devices like fitness trackers, smartwatches, and medical sensors collect real-time health data, enabling continuous monitoring of vital signs, activity levels, and sleep patterns.
- **Telehealth Platforms:** Telemedicine and remote consultation platforms assist virtual visits between patients and healthcare providers, encouraging remote personalized care.

7.3.6.5 EHRs and data integration

- **Interoperable EHR Systems:** Integrated EHRs consolidate patient data from various sources, enabling comprehensive assessment and personalized treatment planning.
- **Health Information Exchange (HIE):** HIE platforms support the secure and seamless exchange of patient information across different healthcare organizations, improving care coordination and fostering better communication among healthcare providers.

7.3.6.6 Drug development and pharmacogenomics

- **Pharmacogenomics:** The study of how an individual's genetic makeup affects their response to medications helps personalize drug selection and dosages.
- **In Silico Drug Screening:** AI-driven simulations help identify potential drug candidates and predict their interactions with biological systems.

7.3.6.7 Surgical planning and robotics

- **Computer-Assisted Surgery:** AI-powered surgical planning tools provide 3D models and simulations for complex surgeries, enhancing precision and safety.
- **Robotic Surgery:** Robots assist surgeons in performing procedures with enhanced accuracy and dexterity.

7.3.6.8 Patient engagement and behavior change

- **Mobile Apps and Health Portals:** Apps and platforms engage patients by providing personalized health information, reminders, and tracking tools for medications and appointments.
- **Behavioral Analytics:** AI analyzes patient behaviors to tailor interventions that encourage healthy habits and compliance with treatment plans.

7.3.6.9 Clinical trials and research

- **Patient Recruitment Tools:** AI identifies suitable candidates for clinical trials based on medical histories, genetic profiles, and demographics.
- **Real-World Evidence Analysis:** AI assesses real-world patient data to generate insights into treatment effectiveness, safety, and outcomes.

7.3.6.10 Population health management

- **Risk Stratification:** AI identifies high-risk populations by analyzing health data, allowing healthcare providers to allocate resources effectively and design preventive interventions.

These tools and techniques collectively enable healthcare providers to offer personalized treatment and care based on individual patient characteristics, medical history, genetics, and preferences.

7.3.7 AI support in healthcare operations and management

AI has the potential to significantly impact healthcare operations and management by streamlining processes, improving efficiency, and enhancing decision-making. Here are several ways AI is being used to support various aspects of healthcare operations and management.

7.3.7.1 Administrative tasks

- **Appointment Scheduling:** AI-powered systems can optimize appointment scheduling, reducing waiting times and ensuring efficient resource allocation.
- **Chatbots and Virtual Assistants:** AI chatbots assist patients in scheduling appointments, answering common queries, and providing basic medical information.

7.3.7.2 Medical billing and claims processing

- **Automated Coding:** AI algorithms can automatically assign appropriate medical codes to diagnoses and procedures, reducing coding errors and improving billing accuracy.
- **Claims Processing:** AI can identify discrepancies and fraudulent claims by analyzing patterns and anomalies in billing data.

7.3.7.3 Supply chain management

- **Inventory Optimization:** AI predicts demand for medical supplies, reducing overstock and stockouts while ensuring the availability of critical items.
- **Predictive Maintenance:** AI monitors equipment and devices to predict maintenance needs, minimizing downtime and disruptions.

7.3.7.4 Patient flow and resource allocation

- **Bed Management:** AI algorithms calculate patient admission and discharge rates, helping hospitals optimize bed utilization and decrease overcrowding.
- **Resource Allocation:** AI serves in allotting resources such as operating rooms, staff, and equipment based on patient requirements and caseloads.

7.3.7.5 Healthcare analytics

- **Population Health Management:** AI analyzes large datasets to identify high-risk patient populations, enabling targeted interventions and preventive measures.
- **Predictive Analytics:** AI models predict disease outbreaks, patient readmissions, and staffing needs based on historical data and trends.
- **Quality Improvement and Patient Safety:** *Early Warning Systems*: AI monitors patient data to identify signs of deterioration or complications, enabling early intervention and reducing adverse events.
- **Root Cause Analysis:** AI helps investigate adverse events by analyzing various factors and identifying contributing factors.
- **Clinical Trials and Research:** *Patient Recruitment*: AI identifies potential clinical trial candidates by analyzing EHRs, genetic profiles, and medical histories.
- **Drug Discovery:** AI assists in drug development by simulating molecular interactions, identifying potential drug candidates, and predicting their properties.

7.3.7.6 Telehealth and remote monitoring

- **Remote Patient Monitoring:** AI-powered devices collect and analyze patient data, providing real-time insights into healthcare providers for the remote management of chronic conditions.
- **Telemedicine:** AI-enhanced telehealth platforms facilitate virtual consultations and remote diagnoses, expanding access to medical care.

7.3.7.7 Risk management and fraud detection

- **Fraud Detection:** AI algorithms analyze claims data to detect fraudulent activities and anomalies, preventing financial losses for healthcare organizations.
- **Clinical Risk Assessment:** AI identifies patterns that indicate potential clinical risks, such as patient deterioration or infection spread.

7.3.7.8 NLP in healthcare documentation

- **EHR Analysis:** NLP techniques extract valuable information from unstructured clinical notes and narratives, aiding clinical decision-making.

By incorporating AI into healthcare operations and management, organizations can optimize workflows, enhance patient care, reduce costs, and improve overall efficiency in a rapidly evolving healthcare landscape.

7.3.8 Tools and techniques used in healthcare operations and management

AI tools and techniques play a crucial role in enhancing healthcare operations and management. They improve efficiency, decision-making, and patient care. Here are some AI tools and techniques commonly used in healthcare operations and management.

7.3.8.1 Predictive analytics

Predictive models use historical data to forecast patient admissions, readmissions, disease outbreaks, and resource needs. This helps hospitals allocate resources effectively and plan for patient influxes.

7.3.8.2 Natural language processing

NLP techniques analyze and extract insights from unstructured clinical notes, medical literature, and patient feedback. This aids in clinical decision support, research, and sentiment analysis.

7.3.8.3 Robotic process automation

Robotic process automation automates repetitive administrative tasks like data entry, claims processing, and appointment scheduling, reducing human errors and freeing up staff for more complex tasks.

7.3.8.4 Chatbots and virtual assistants

AI-powered chatbots provide quick responses to patient inquiries, appointment scheduling, and basic medical information, improving patient engagement and reducing staff workload.

7.3.8.5 Telehealth platforms

Telemedicine tools equipped with AI capabilities enable virtual consultations, remote diagnostics, and real-time monitoring of patients, expanding access to care.

7.3.8.6 Supply chain optimization

AI optimizes inventory management by predicting demand for medical supplies and equipment, preventing shortages, and reducing excess stock.

7.3.8.7 Patient flow management

AI predicts patient admission and discharge rates, helping hospitals manage bed availability, reduce wait times, and enhance patient flow.

7.3.8.8 Clinical decision support systems

AI analyzes patient data, medical literature, and best practices to provide evidence-based treatment recommendations to clinicians.

7.3.8.9 Healthcare analytics platforms

AI-driven analytics platforms aggregate and analyze data from various sources, allowing administrators to monitor key performance indicators, patient outcomes, and operational efficiency.

7.3.8.10 Smart health records

AI-powered EHR systems use machine learning to identify trends, anomalies, and potential risks within patient data, improving care quality and safety.

7.3.8.11 Fraud detection and prevention

AI algorithms analyze claims data and patterns to identify potential fraudulent activities, reducing financial losses for healthcare organizations.

7.3.8.12 Image and diagnostic analysis

AI-powered diagnostic tools analyze medical images (e.g., X-rays, MRIs) to detect anomalies and assist radiologists in accurate diagnoses.

7.3.8.13 Remote monitoring devices

Wearable devices with AI capabilities monitor patient vitals and transmit data to healthcare providers, enabling proactive interventions and remote management of chronic conditions.

7.3.8.14 Population health management

AI analyzes large datasets to identify high-risk patient populations, enabling healthcare organizations to implement preventive measures and targeted interventions.

7.3.8.15 Operational efficiency optimization

AI analyzes hospital workflows and resource allocation to identify bottlenecks, inefficiencies, and opportunities for improvement.

The integration of these AI tools and techniques into healthcare operations and management results in streamlined processes, improved patient outcomes, reduced costs, and enhanced overall quality of care.

7.4 RESULT AND DISCUSSION

Table 7.1 shows the application of AI in healthcare and tools and techniques used to find the outcome of this field.

7.5 CONCLUSION AND FUTURE WORK

In conclusion, the integration of AI into healthcare is reshaping the industry in profound ways. The applications mentioned above represent just a fraction of the potential AI holds in improving healthcare delivery. As AI technologies continue to advance, the following future directions and considerations will be crucial. *Ethical and Regulatory Frameworks*: The development and deployment of AI in healthcare must adhere to robust ethical guidelines and regulatory standards. Patient privacy, data security, and transparency in AI algorithms are paramount concerns that need ongoing attention. *Data Quality and Interoperability*: High-quality and interoperable healthcare data are essential for AI systems to provide accurate insights. Efforts to standardize and share healthcare data across systems and institutions are critical. *AI Explainability*: As AI algorithms become more complex, ensuring that their decisions are explainable to clinicians

Table 7.1 Application of AI in Healthcare and Its Outcome

Application of AI in Healthcare	Technology Included	Prediction	Tools and Techniques Used	Outcome
Medical imaging analysis	Image recognition and classification	Types of tumors, fractures, or lesions in X-rays or MRIs.	Convolutional neural networks, image registration, image fusion, random forests and decision trees, texture analysis	AI algorithms can automatically identify and classify structures and anomalies in medical images.
	Segmentation and anatomical mapping	AI can segment medical images to isolate specific organs, tissues, or structures.	Segmentation algorithms	This aids in quantifying abnormalities and measuring their dimensions.
Drug discovery and development	Virtual screening and compound design	AI models analyze vast chemical libraries to predict which molecules are likely to interact with specific disease targets.	Data mining and knowledge discovery	This helps identify potential drug candidates more efficiently.
	Predicting drug–target interactions	AI techniques predict interactions between drug molecules and biological targets.	QSAR, machine learning models: Algorithms such as random forests, support vector machines, and deep learning models predict interactions between drugs and target proteins based on molecular features.	This aids in understanding drug mechanisms and identifying potential side effects.
Personalized treatment and care	Medical diagnostics and imaging	AI algorithms can analyze medical images such as X-rays, MRIs, and CT scans to assist in early and accurate diagnosis of diseases, patterns, and anomalies.	CAD, imaging biomarkers	This identifies abnormalities, assisting radiologists and clinicians in accurate diagnosis.

(Continued)

Table 7.1 (Continued) Application of AI in Healthcare and Its Outcome

Application of AI in Healthcare	Technology Included	Prediction	Tools and Techniques Used	Outcome
	Genomic analysis	AI can analyze genetic data to identify genetic variations that might be associated with certain diseases or conditions.	Next-generation sequencing, bioinformatics	This identifies genetic variations that could impact disease risk and treatment response.
	Remote monitoring and telehealth	AI-powered devices and wearable technologies can continuously monitor patients' vital signs and health metrics.	Fitness trackers, smartwatches, and medical sensors, HER, HIE	This data can be used to provide real-time insights into both patients and healthcare providers, enabling early intervention and personalized adjustments to treatment plans.
	Surgical planning and robotics	AI-assisted surgical tools and robots enhance the precision and accuracy of surgical procedures.	Computer-assisted surgery, robotic surgery	They can adapt to the patient's anatomy and provide surgeons with real-time feedback during operations.
Healthcare operations and management	Administrative tasks	AI-powered systems optimize appointment scheduling, reduce waiting times, and improve resource allocation.	Predictive analytics, natural language processing, robotic process automation	This reduces waiting times, ensuring efficient resource allocation.
	Chatbots and virtual assistants	AI chatbots handle scheduling, answer common queries, and provide basic medical information, improving efficiency and reducing administrative burden.	Chatbots and virtual assistants	This reduces waiting times, ensuring efficient resource allocation.
	Supply chain management	AI predicts demand for medical supplies and reduces overstock and stockouts while ensuring the availability of critical items	Supply chain management	AI optimizes inventory management by predicting demand for medical supplies and equipment, preventing shortages, and reducing excess stock.

is essential for gaining trust and acceptance. Explainable AI research is vital for this purpose. *Integration with Clinical Workflow*: AI solutions need to seamlessly integrate into the clinical workflow to maximize their utility. User-friendly interfaces and effective training for healthcare professionals are essential. *Continual Learning and Validation*: AI models must be continuously updated and validated using real-world data to ensure their ongoing accuracy and relevance. *Expanding to Underserved Areas*: AI can be particularly beneficial in underserved or remote areas where access to healthcare expertise is limited. Efforts to make AI-driven healthcare accessible globally should be a priority. *AI in Public Health*: Expanding the use of AI for public health surveillance, outbreak prediction, and epidemiological studies can help mitigate global health crises. *Collaboration between AI and Healthcare Professionals*: AI should be viewed as a tool to enhance the capabilities of healthcare professionals, not replace them. Collaboration and mutual understanding between AI developers and healthcare practitioners are crucial.

REFERENCES

1. Silver, D., Schrittwieser, J., Simonyan, K., Antonoglou, I., Huang, A., Guez, A., ... & Hassabis, D. (2017). Mastering chess and shogi by self-play with a general reinforcement learning algorithm. *Nature*, 550(7676), 354–359.
2. Silver, D., Huang, A., Maddison, C. J., Guez, A., Sifre, L., Van Den Driessche, G., ... & Hassabis, D. (2016). Mastering the game of Go with deep neural networks and tree search. *Nature*, 529(7587), 484–489.
3. LeCun, Y., Bengio, Y., & Hinton, G. (2015). Deep learning. *Nature*, 521(7553), 436–444.
4. Esteva, A., Kuprel, B., Novoa, R. A., Ko, J., Swetter, S. M., Blau, H. M., & Thrun, S. (2017). Dermatologist-level classification of skin cancer with deep neural networks. *Nature*, 542(7639), 115–118.
5. Litjens, G., Kooi, T., Bejnordi, B. E., Setio, A. A. A., Ciompi, F., Ghafoorian, M., ... & Sanchez, C. I. (2017). A survey on deep learning in medical image analysis. *Medical image analysis*, 42, 60–88.
6. Zhu, Z., Zhang, Q., Wang, W., & Yuille, A. L. (2019). Deep learning in visual computing. *Journal of Computer Science and Technology*, 34(1), 7–24.
7. McKinney, S. M., Sieniek, M., Godbole, V., Godwin, J., Antropova, N., Ashrafian, H., ... & Reicher, J. J. (2020). International evaluation of an AI system for breast cancer screening. *Nature*, 577(7788), 89–94.
8. Bzdok, D., & Meyer-Lindenberg, A. (2018). Machine learning for precision psychiatry: opportunities and challenges. *Biological Psychiatry: Cognitive Neuroscience and Neuroimaging*, 3(3), 223–230.
9. Esteva, A., Robicquet, A., Ramsundar, B., Kuleshov, V., DePristo, M., Chou, K., ... & Corrado, G. (2019). A guide to deep learning in healthcare. *Nature Medicine*, 25(1), 24–29.
10. Chen, J. H., Asch, S. M., & Machine, L. (2017). The role of virtual teams in the healthcare workplace. *Human Resources for Health*, 15(1), 1–5.

11. Rajkomar, A., Oren, E., Chen, K., Dai, A. M., Hajaj, N., Hardt, M., ... & Zhang, M. (2018). Scalable and accurate deep learning with electronic health records. *NPJ Digital Medicine*, 1(1), 1–10.
12. Beam, A. L., & Kohane, I. S. (2018). Big data and machine learning in health care. *JAMA*, 319(13), 1317–1318.
13. He, K., Zhang, X., Ren, S., & Sun, J. (2016). Deep residual learning for image recognition. In *Proceedings of the IEEE Conference on Computer Vision and Pattern Recognition (CVPR)*, United States (pp. 770–778).
14. Meng, Z., Meng, L., & Tomiyama, H. (2021). Pneumonia diagnosis on chest X-rays with machine learning. *Procedia Computer Science*, 187, 42–51, China.
15. Tan, M., & Le, Q. V. (2019). EfficientNet: Rethinking model scaling for convolutional neural networks. In *Proceedings of the 36th International Conference on Machine Learning (ICML)* (Vol. 97, pp. 6105–6114). USA: California.
16. Abhilasha Rajput, Ajay Singh, Pushpneel Verma (2023). Study on heart disease predication using artificial intelligence. *IJCRT*, 11 (6). ISSN: 2320-2882.
17. Mann, J. A., Dvorak, M., Wright, T. M., & Windsor, R. E. (2012). Computer-assisted robotic total hip arthroplasty compared with manual total hip arthroplasty. JBJS, 94(3), 264–273.
18. Behara, K., Bhero, E., Agee, J. T., & Gonela, V. (2022). Artificial intelligence in medical diagnostics: A review from a South African context. *Scientific African*, 17, e01360.

## Chapter 8

Revolutionising chronic healthcare management

Integrating AI and IoT for personalised patient care

Gnanajeyaraman Rajaram, Deepa Rajamanickam, Meenalochini Jeganathan, and R. S. Archana Vishveswari

8.1 INTRODUCTION

Healthcare has experienced transformational changes as a result of the quick development of technology [1]. A new age known as the Artificial Intelligence of Things (AIoT), which has enormous promise for altering the landscape of healthcare, has begun with the merging of artificial intelligence (AI) with the Internet of Things (IoT) [2]. This convergence offers cutting-edge answers to enduring problems in health monitoring and diagnosis, marking a substantial shift from conventional healthcare techniques. The AIoT paradigm is set to revolutionise how healthcare is given, experienced, and optimised by fusing the strength of AI's data processing and predictive powers with IoT's seamless connectivity and data collecting [3]. AIoT in healthcare tackles challenges by enhancing access in remote areas through telemedicine and mobile health solutions. It enables continuous monitoring using wearable devices and remote sensors, improving patient care by providing real-time health data analysis and timely interventions, ultimately leading to better health outcomes and reduced healthcare disparities.

Figure 8.1 illustrates the AIoT ecosystem in healthcare, demonstrating how AI and IoT technologies are seamlessly integrated. It demonstrates the collaborative collection, transmission, and analysis of health data via

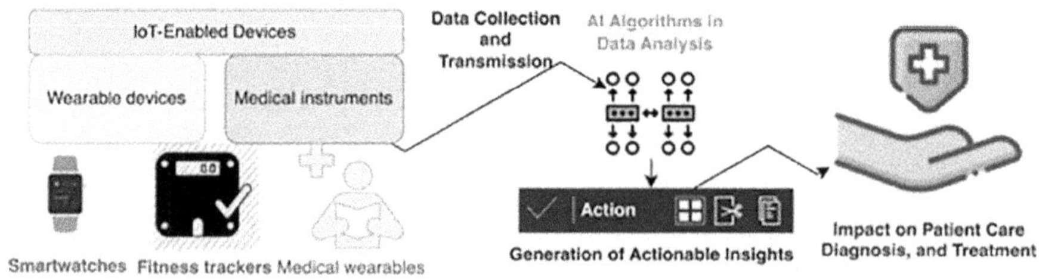

Figure 8.1 AIoT ecosystem in healthcare.

DOI: 10.1201/9781032618173-10

a network of linked wearables, medical devices, and other IoT-enabled devices. The image emphasises how important AI algorithms are in digesting the gathered data to produce insightful knowledge. The quality of patient care is subsequently improved, correct diagnoses are made possible, and treatment plans are improved thanks to these insights.

8.1.1 The synergy of AI and IoT in healthcare: a paradigm shift

Beyond incremental advances, the merging of AI and IoT technology signifies a fundamental leap in healthcare. Traditional healthcare models frequently relied on patients visiting facilities on an as-needed basis to have their ailments evaluated and treated for brief periods of time [4]. Although useful, this technique has problems portraying the complex and changing character of people's health. By enabling continuous, real-time monitoring of patients' health indicators and vitals, the AIoT paradigm addresses these limitations. A thorough, all-encompassing picture of patients' health statuses is created with the help of smart medical devices, wearable gadgets with embedded sensors, and connected medical records. The analysis of predicted health deteriorations serves as an example of how combining AI and IoT improves healthcare service. It anticipates possible health deterioration by combining AI algorithms with IoT sensor data from wearable devices. This allows for early interventions and individualised care plans, which improves patient outcomes.

8.1.2 Challenges in health monitoring and diagnosis: the AIoT solution

Healthcare has struggled for a very long time with issues including timely health monitoring, precise diagnosis, and efficient treatment. Patients' ability to receive prompt care has been hampered by the lack of access to healthcare facilities, especially in remote or underdeveloped locations [5]. A proactive and personalised approach to healthcare is also required because of the variation in patient conditions and responses to treatments. It is impossible to overstate how the AIoT is emerging as a solution to these problems. Healthcare systems are given the ability to collect, analyse, and interpret data in previously unthinkable ways by utilising the capabilities of AI and IoT.

By using wearable technology to get over conventional system constraints, AIoT tackles the problems associated with health monitoring. These gadgets' AI algorithm integration allows for continuous, remote vital sign monitoring, which helps with the early identification of health problems. Proactively improving patient care results in prompt treatments and better health outcomes.

8.1.3 Improved patient care: the heart of the AIoT revolution

The AIoT has the ability to completely transform patient care, which is one of its most significant contributions to healthcare. Reactive interventions—treatment given after symptoms or illnesses manifested—were common in traditional healthcare. Proactive and preventative healthcare is now the main focus thanks to the AIoT. Continuous health monitoring is made possible by wearable technology, which has sensors that track heart rate, blood pressure, activity levels, and even sleep patterns [6]. Healthcare practitioners receive this real-time data, which enables them to spot early indications of anomalies or variations from typical patterns. Think about a patient who wears an AIoT-enabled smartwatch and has a history of cardiac problems. The watch continuously tracks the patient's heart rate and rhythm, looking for any abnormalities in the information. An alert is generated for the patient and their healthcare physician if the AI algorithms find any odd patterns that could point to an impending cardiac attack. This prompt action could avert a potentially fatal circumstance, demonstrating the potential of AIoT to enhance patient care. The authors [7] cite several studies and statistics to demonstrate the effectiveness of AIoT in improving patient care, such as reducing hospital readmissions, enhancing diagnosis accuracy, and optimising treatment outcomes.

8.1.4 Cost reduction by taking preventative action

The AIoT might drastically lower healthcare expenses in addition to having an impact on patient care. In addition to improving patient outcomes, the shift from reactive to proactive treatment reduces the need for costly interventions and hospital stays. Early detection of health disorders enables healthcare professionals to carry out less invasive and expensive procedures, halting the development of illnesses that could require substantial treatment [8]. Additionally, the AIoT's predictive powers might help with resource management and allocation in healthcare systems. Hospitals can optimise their resources, lowering inefficiencies and costs, by making accurate predictions of patient intake, disease outbreaks, or changes in health demand. By reducing downtime and ensuring optimal operation, the capacity to anticipate equipment maintenance requirements based on data gathered from IoT-connected medical devices can also help to save money. AIoT in healthcare reduces costs through remote patient monitoring, predictive analytics for disease prevention, smart health monitoring, medication adherence tools, and telemedicine. By detecting issues early, personalising care, and enabling remote interventions, it lowers hospital visits, complications, and treatment expenses, enhancing both patient outcomes and cost efficiency.

8.1.5 Improved patient care: the heart of the AIoT revolution

The AIoT has an impact on people's general well-being and empowerment that goes beyond only medical therapies. People become more conscious of their own health conditions as a result of ongoing health monitoring. With this newfound understanding, they are better equipped to take preventative steps to safeguard their well-being, whether through dietary adjustments, medication adherence, or routine check-ups with medical professionals. An AIoT-enabled glucose monitoring device, for instance, can measure a person's blood sugar levels and provide real-time feedback if they are managing diabetes. Based on the data gathered, the companion AI-powered app can provide tailored food suggestions and modifications to insulin dosage. This not only enhances the person's health but also cultivates a feeling of ownership and control over their well-being. The AIoT paradigm, a disruptive force that reimagines how healthcare is provided and experienced, was created as a result of the combination of AI and IoT technologies in healthcare. The AIoT has the potential to revolutionise patient care, cut costs, and improve overall well-being because of its capacity to enable continuous health monitoring, offer predictive insights, and encourage proactive care. Healthcare systems, practitioners, and patients stand to gain from a data-driven, individualised, and effective approach to health monitoring and diagnosis as this convergence gets hold. The potential is clear, but it's crucial to negotiate the ethical, privacy, and legal issues that arise from the enormous volume of data generated and processed inside the AIoT ecosystem. The transformational influence of the AIoT on healthcare remains a promise and a responsibility as we work through these difficulties—a force for good that necessitates careful stewardship and creativity.

8.2 WEARABLE DEVICES AND REMOTE PATIENT MONITORING: PIONEERING PERSONALISED HEALTHCARE THROUGH AIoT

The fusion of technology and medicine has produced ground-breaking technologies that are changing the way patient care is provided and experienced in the ever-evolving world of healthcare. The incorporation of wearable technology and remote patient monitoring into the AIoT framework is one such breakthrough that has attracted substantial attention and shows great promise [8]. These wearables are revolutionising healthcare by enabling continuous and real-time health data collection, transmission, and analysis. They are outfitted with cutting-edge sensors and seamlessly integrated into the IoT ecosystem. This change is enabling people to actively manage their own well-being while simultaneously improving healthcare professionals' ability to keep an eye on patients' problems.

Limitations of wearable devices include accuracy concerns due to sensor limitations and user compliance issues like inconsistent wear. Overcoming these challenges involves improving sensor technology for better accuracy, enhancing device comfort and aesthetics to encourage consistent wear, and integrating AI algorithms to compensate for data inaccuracies, ensuring reliable health monitoring and user engagement.

8.2.1 The emergence of wearable devices in healthcare

Wearable technology has expanded beyond its original use in consumer electronics and fitness tracking to become a vital component of contemporary healthcare. These gadgets have a variety of sensors that can track different physiological characteristics, ranging from smartwatches and fitness bands to medical-grade wearable patches [9,10]. A central platform may now wirelessly receive continuous measurements of heart rate, temperature, blood pressure, respiration rate, activity levels, and even electrocardiograms for analysis.

Figure 8.2 illustrates the idea of wearable technology and remote patient monitoring in the context of AIoT in healthcare. It features a variety of wearable gadgets with sensors, including smartwatches, activity trackers, and medical wearables. These devices collect data that is sent to healthcare professionals for ongoing monitoring, allowing for the early identification of irregularities and prompt intervention.

8.2.2 The benefits of ongoing surveillance

In the past, clinical settings have relied significantly on episodic patient visits when conditions are assessed over the course of brief interactions. However, these "snapshots" of health sometimes fall short of capturing the dynamic character of health situations, especially when little changes may be a sign of more serious underlying problems. In order to close this gap, wearable technology, which is a component of the AIoT ecosystem, makes it possible to continuously monitor patients' physiological data. This continuous data collection gives healthcare providers a thorough insight into

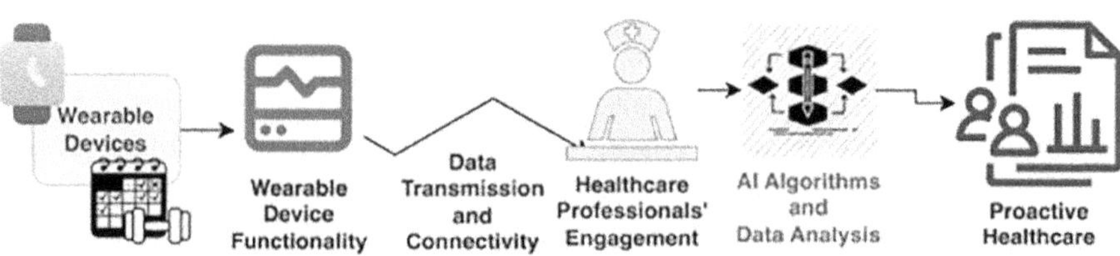

Figure 8.2 Wearable devices and remote patient monitoring.

patients' health conditions in real time, enabling them to spot anomalies, trends, and departures from normal patterns [11].

8.2.3 Data insights to empower healthcare professionals

Continuous health monitoring is made more useful by the AIoT paradigm's incorporation of wearable technology. When subjected to AI-driven analysis, the data acquired from wearables is transformed into more than simply raw data and becomes an effective tool for healthcare practitioners [12]. Massive amounts of data can be sorted through machine learning algorithms, which can then spot patterns, correlations, and anomalies that might not be seen using more conventional approaches. For instance, AI systems can link changes in sleep patterns to the onset of a mood problem or they can identify minor variations in heart rate variability that may signal an approaching cardiac issue.

8.2.4 Beyond clinical setting boundaries: remote patient monitoring

The idea of remote patient monitoring is one of wearable technology's most important contributions to the AIoT framework. In the traditional patient care approach, monitoring and evaluation frequently demanded that patients be physically present in clinical settings [13]. This restriction has been overcome by the widespread use of wearable technology and its incorporation into the AIoT ecosystem. Outside of hospitals or clinics, patients can now be remotely monitored. Imagine an older patient with a chronic ailment wearing a wearable that tracks their vital signs and level of activity all the time. A healthcare provider's dashboard receives this data in real time. Without the patient having to travel, the practitioner can evaluate the patient's state, enabling prompt actions in the event of anomalies or deterioration. In addition to enhancing patient comfort and quality of life, remote patient monitoring eases the burden on healthcare institutions, which is crucial in resource constrained situations or during public health emergencies.

When it comes to early diagnosis and timely interventions, the capability of monitoring patients remotely and in real time is especially important. Wearable technology has the potential to detect minute changes in health markers that could signal the beginning of a medical problem. For instance, a wearable gadget monitoring a person's blood glucose levels might spot a tendency towards hyperglycaemia, alerting medical professionals to take action before the issue worsens and turns into an emergency. By addressing problems before they become critical, early diagnosis not only enhances patient outcomes but also lessens the burden on healthcare resources.

Challenges and considerations about the integration of wearable technology and remote patient monitoring within the AIoT ecosystem have enormous potential, but there are still certain difficulties. Wearable technology gathers private health data that must be protected from unauthorised access and breaches, making data security and patient privacy top priorities. Furthermore, it's crucial to guarantee the reliability and accuracy of the data gathered by these devices to make wise therapeutic judgements. The issue of interoperability is another one that needs to be resolved. For effective remote patient monitoring, the healthcare ecosystem includes a wide range of devices from diverse manufacturers. Assuring seamless integration and data exchange between various wearables and platforms is therefore essential. There are currently attempts being made to standardise to provide frameworks and protocols that allow for data sharing and interoperability.

The AIoT paradigm's incorporation of wearable technology and remote patient monitoring heralds a seismic upheaval in healthcare. By providing healthcare professionals with real-time data insights, this convergence enables them to make wise judgements and proactively intervene. Wearable technology enables continuous monitoring, which goes beyond the constraints of episodic healthcare and provides a full and dynamic perspective of patients' health conditions. The potential for personalised, data-driven healthcare is limitless as technology develops and AI algorithms get more complex. A new era in healthcare monitoring—one that is patient-centric, proactive, and profoundly informed by the power of data and technology—is about to begin with the marriage of wearable devices and the AIoT.

Use Case: Wearable Technology and Remote Patient Monitoring: Personalised Healthcare Pioneering with the IoT

Modern medical knowledge and cutting-edge technology have spawned a revolution in the dynamic world of healthcare that is fundamentally altering patient care. The combination of wearable technology and remote patient monitoring falls under the umbrella of the AIoT, and it is one of these paradigm-shifting developments that has the most promise. These wearable wonders are enabling continuous real-time health data gathering, transmission, and in-depth analysis, ushering in a paradigm shift in healthcare. They are seamlessly woven into the IoT fabric. Health management is being revolutionised by the convergence of AIoT, which is driving healthcare practitioners towards more individualised, proactive, and patient-centred care.

8.2.4.1 The genesis of wearables in healthcare

Wearable technology has progressed from being seen as little more than fashion items to being an essential part of the healthcare industry. These gadgets, which range from stylish smartwatches to cutting-edge wearable medical patches, are furnished with a variety of sensors that go beyond conventional health measurements. Heart rate, body temperature, blood pressure, respiration rate, activity levels, and even intricate electrocardiograms are among the range of data that can be recorded. Wearables offer a chance to revolutionise the monitoring of health issues by gathering this data in an unobtrusive and non-invasive manner.

8.2.4.2 The power of continuous monitoring unleashed

In the past, healthcare was provided through irregular clinic visits that merely provided fragmented glimpses of patient health. However, the episodic method frequently falls short of capturing the dynamic nature of medical disorders. The AIoT ecosystem's wearable technology is intended to close this gap through ongoing health monitoring. This continuous flow of data allows healthcare practitioners to have an uninterrupted picture of patients' health states and trends, enabling a thorough and real-time understanding of patients' well-being.

8.2.4.3 Improved healthcare using AI insights

While wearables' ability to collect data continuously is a step forward, it is when this data is used by AI-driven analytics that a true revolution takes place. When subjected to cutting-edge machine learning algorithms, the collected data transforms from simple data to a potent tool. Large data sets can be combed through by these algorithms, which can also spot abnormalities, subtle correlations, and detailed patterns that can escape traditional research. AI systems, for example, can identify minute variations in heart rate variability that portend cardiac problems or link changes in sleep patterns to the onset of psychological diseases.

8.2.4.4 Remote patient monitoring is a redefinition of healthcare delivery

The idea of remote patient monitoring is perhaps the most revolutionary component of wearables inside the AIoT framework. For patient assessment and monitoring, traditional healthcare models require physical presence. This dynamic changes as wearables become more prevalent in the AIoT environment and enable remote patient monitoring outside of institutional settings for healthcare.

Consider an elderly patient who uses a wearable device to manage a chronic disease. This device continuously tracks activities and vital signs, sending real-time information to a healthcare provider's dashboard. This arrangement makes it possible for prompt interventions to be made without the patient having to travel, improving patient comfort and expediting medical procedures. In addition to ensuring improved patient experiences, remote patient monitoring is essential for easing the pressure on healthcare facilities when resources are scarce.

8.2.4.5 Early detection: a decisive development

Continuous and remote monitoring becomes extremely important in early diagnosis and prompt treatments thanks to AI-driven analytics. Wearables have the potential to detect minute changes in health parameters, which frequently signal impending medical problems. A wearable gadget that tracks blood sugar levels, for instance, may foresee hyperglycaemic tendencies and enable healthcare professionals to take preventative action. By catching possible problems early on, this early identification not only improves patient outcomes but also relieves demand for healthcare resources. Even while wearables and remote patient monitoring fall within the AIoT umbrella, they come with difficulties. Since wearables capture private health information that needs to be protected to the highest standard, data security and patient privacy are of utmost importance. For making educated decisions, maintaining data accuracy and reliability is equally important. Because there are so many different types of healthcare technologies, achieving interoperability—the smooth integration of devices—can be difficult. Ongoing standardisation initiatives aim to create uniform protocols that guarantee data sharing and interoperability.

A new age in healthcare is being ushered in by the combination of wearable technology and remote patient monitoring made possible by the AIoT. The real-time insights provided by this connection enable healthcare practitioners to take proactive action and make well-informed judgements. Patient management that is comprehensive, ongoing, and AI-powered is replacing the era of sporadic healthcare. The potential for personalised, data-driven healthcare is limitless as technology and AI algorithms advance. A new era that is patient-centric, predictive, and powered by the symbiosis of data and technology is heralded by the marriage of wearables and AIoT. It's a time when the patient is put first, allowing medical professionals to give solutions that are both successful and individualised.

Use Case Sample Output:

```
Enter patient ID to monitor (or 'exit' to quit): P1
Patient P1 data collected: {'heart_rate': 95, 'temperature': 36.75267745228145, 'blood_pressure': '128/82'}
Enter patient ID to monitor (or 'exit' to quit): P2
Patient P2 data collected: {'heart_rate': 72, 'temperature': 36.63403797078335, 'blood_pressure': '98/70'}
Enter patient ID to monitor (or 'exit' to quit): P3
Patient P3 data collected: {'heart_rate': 93, 'temperature': 37.16462019010231, 'blood_pressure': '131/80'}
Enter patient ID to monitor (or 'exit' to quit): P4
Patient P4 data collected: {'heart_rate': 94, 'temperature': 36.75712691879364, 'blood_pressure': '99/76'}
```

Enter patient ID to monitor (or 'exit' to quit): P5
Patient P5 data collected: {'heart_rate': 71, 'temperature': 37.278330034905856, 'blood_pressure': '135/74'}
Enter patient ID to monitor (or 'exit' to quit): P6
Patient P6 data collected: {'heart_rate': 93, 'temperature': 36.83004652386007, 'blood_pressure': '136/64'}
Enter patient ID to monitor (or 'exit' to quit): P7
Patient P7 data collected: {'heart_rate': 81, 'temperature': 36.557746864647086, 'blood_pressure': '110/89'}
Enter patient ID to monitor (or 'exit' to quit): P8
Patient P8 data collected: {'heart_rate': 88, 'temperature': 36.695475909247556, 'blood_pressure': '136/72'}
Enter patient ID to monitor (or 'exit' to quit): P9
Patient P9 data collected: {'heart_rate': 93, 'temperature': 36.70423922107684, 'blood_pressure': '92/73'}
Enter patient ID to monitor (or 'exit' to quit): exit

Figure 8.3 shows patients' health data gathered in a remote patient monitoring scenario as a visual representation. The graph displays each patient's temperature and heart rate readings, providing details about their current state of health. The x-axis of the graph is organised by patient IDs, and each patient is represented by two sets of vertical bars, one for their heart rate and the other for their body temperature.

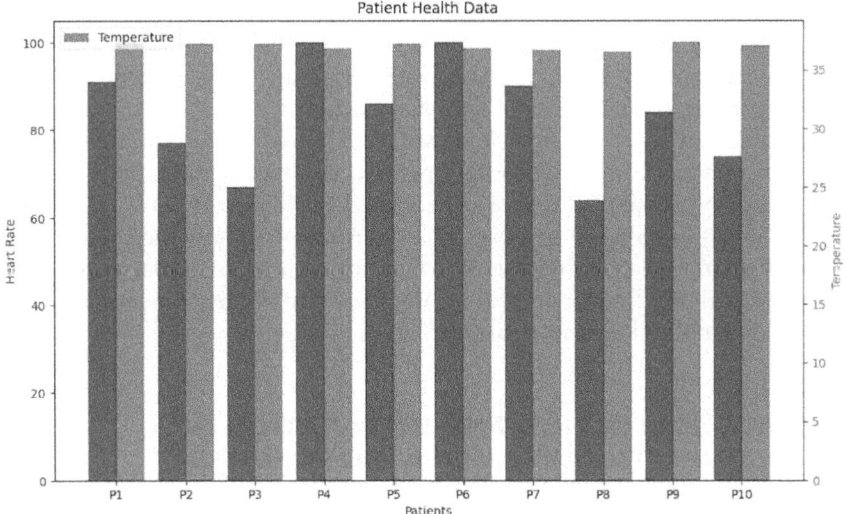

Figure 8.3 Health data visualisation for patients.

8.2.4.6 Blue bars: heart rate

The patient's heart rate values are shown by the blue bars. Each blue bar's height reflects the patient's specific heart rate value. Lower blue bars denote lower heart rates, whereas larger blue bars denote higher heart rates. The left y-axis shows the heart rate readings, while the blue bars are marked "Heart Rate."

8.2.4.7 Red bars: temperature

The patients' body temperatures are represented by the red bars. Each red bar's height corresponds to the patient's specific temperature measurement. The right y-axis shows the temperature data, while the red bars are marked "Temperature."

- **Patient IDs:** The patient ID, which is shown on the x-axis, is used to uniquely identify each patient. The patient IDs are shown aligned beneath their corresponding bars on the "Patients" x-axis.
- The graph offers a rapid visual comparison of temperature and heart rate readings for various patients. Healthcare workers can rapidly spot differences in patients' heart rates and temperatures by watching the heights of the blue and red bars. The ability to spot patterns, trends, or probable abnormalities in health data is made possible by this visualisation.
- **Decision-Making and Understanding:** Healthcare professionals can use the graph as a useful tool to simultaneously track the health conditions of several patients. The visual display of medical data makes it easier to spot patients with unusual heart rates or body temperatures, allowing for rapid, essential actions. Additionally, the graph's patterns and trends might help decision-makers make well-informed choices regarding the best patient care approaches.
- **Source of the Data and Analysis:** Patients who use wearable devices for remote health monitoring provided the data depicted in the graph. After processing, the data is presented on the graph as vertical bars. The data visualisation serves as an illustration of how the combination of wearable technology, remote monitoring, and data analysis can give medical practitioners useful information that will enable proactive and individualised patient treatment.

8.3 DATA COLLECTION AND INTEGRATION IN AIoT HEALTHCARE: UNVEILING INSIGHTS FOR DIAGNOSIS AND TREATMENT

The blending of AI and the IoT has accompanied a new era of data-driven patient care in the changing environment of contemporary healthcare [14]. The seamless integration of data gathered from a variety of sources inside

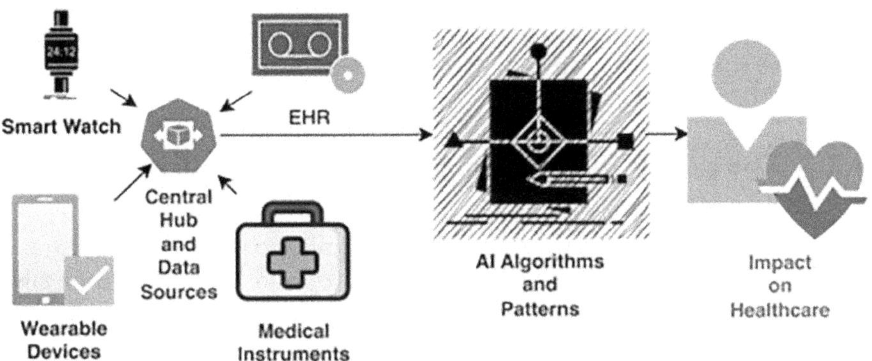

Figure 8.4 Data collection and integration in AIoT healthcare.

the AIoT architecture is at the core of this change. The collection, aggregation, and analysis of health-related data have been completely transformed by this convergence, opening the door to improved methods of diagnosis and therapy [15]. Healthcare professionals can access a wealth of insights that were previously hidden in the massive sea of information by utilising the interconnection of the IoT- and AI-driven algorithms.

The procedure for data gathering and integration in AIoT healthcare is shown in Figure 8.4. It demonstrates how data is compiled from various sources, such as social media, wearable technology, and medical equipment. These data are analysed by AI-driven algorithms to uncover patterns, correlations, and useful information for prognosis and treatment.

Diverse formats and protocols are the root cause of data compatibility problems in AIoT healthcare. Encouraging device interoperability and standardising data formats like HL7 FHIR are essential. Consistent data interpretation is ensured by integrating ontologies with semantic interoperability. Data exchange can be made smoother by using APIs and middleware technologies. Blockchain improves data sharing's security and openness. Heterogeneous data is harmonised for proper analysis using AI-driven techniques. Real-time data processing is facilitated by the development of edge computing solutions, cloud-based integration platforms, and interoperable systems. These tactics facilitate seamless data gathering and integration, revealing insightful information for diagnosis and treatment and encouraging cooperation amongst various healthcare ecosystems.

8.3.1 The IoT ecosystem: a network of data streams

A huge network of networked gadgets and sensors, including those used in smart homes and industrial automation, define the IoT ecosystem. This environment takes on a special allure in the healthcare setting [16]. A rich tapestry of health-related data is produced by wearable technology, medical equipment, electronic health records (EHRs), home monitoring systems, and even social media platforms. This information covers a wide range of factors, such as physiological measurements (heart rate, blood pressure, and temperature), dietary patterns, medication adherence, and even patient-reported outcomes.

8.3.2 The power of combining various data sources

When these various data sources are combined and utilised as a whole, the AIoT's entire potential for healthcare applications becomes apparent. In the past, patient health data was fragmented and kept in separate silos by different healthcare organisations and systems [17]. These hurdles are removed by AIoT, enabling the combination of wearable data with prior medical records, lab findings, and lifestyle data. Informed decisions can be made by healthcare practitioners based on a thorough understanding of each patient's health trajectory thanks to the resulting data mosaic, which offers a comprehensive perspective of patients' health profiles.

8.3.3 Analytics driven by AI: from data to insights

No matter how varied or substantial, raw data is useful only when it is converted into insights that can be put to use. This is where the IoT and AI work best together. The vast amounts of data created by the AIoT ecosystem can be processed and analysed by AI-driven algorithms to uncover patterns, correlations, and insights that may be invisible to human observers. Deep learning models, neural networks, and machine learning algorithms can uncover patterns that reveal complicated linkages in the data, allowing for the drawing of insightful conclusions. Take into account, for illustration, a situation in which data from wearables, EHRs, and patient-reported outcomes are pooled and put via AI-driven analysis. The algorithms may find a link between a patient's level of physical activity and the development of a chronic ailment. With this knowledge, healthcare providers can create individualised treatment regimens that emphasise greater physical activity, possibly halting the progression of the ailment. By moving beyond generalised techniques and customising solutions to each patient's particular circumstances, such discoveries have the potential to revolutionise treatment strategies.

8.3.4 From insights to proactive healthcare

A paradigm change towards proactive healthcare may be enabled by the insights gained from AI-driven analysis of compiled health data. Healthcare interventions have historically frequently been reactive, responding to symptoms or emergencies as they developed. Healthcare practitioners may now identify potential health issues before they formally present symptoms thanks to the AIoT, which is powered by data-driven insights. Practitioners can intervene early, minimising problems and improving patient outcomes, by spotting deviations from normal patterns and associating them with certain health issues. Consider a patient who is at risk for a cardiac attack because of their age, family history, and way of life. Subtle variations in the patient's heart rate variability are found by continuous monitoring and AI analysis; these alterations can go unreported during normal examinations. The healthcare personnel are prompted to begin actions to stop the event from happening by the AI algorithms that identify these patterns as warning signs of oncoming heart trouble. By avoiding emergency procedures, this proactive approach to healthcare not only saves lives but also lowers healthcare expenses.

8.3.5 Privacy and ethical considerations for data

Despite the incredible potential of AIoT-powered healthcare, it's critical to negotiate the ethical questions and data privacy issues that come with the collection and analysis of private health information. The importance of patient permission, data protection, and adherence to privacy laws cannot be overstated. The successful adoption of AIoT technologies in healthcare depends on finding a balance between the potential advantages and the preservation of people's rights.

A new age in healthcare diagnostics and treatment has begun with the merging of AI-driven analysis with IoT-generated health data. The IoT ecosystem's connectivity enables the collection of various data sources, which results in the creation of a thorough picture of patients' health profiles. This data is transformed from raw data to actionable insights by AI-driven algorithms, exposing correlations, trends, and patterns that direct diagnosis and treatment choices. Patient care could undergo a transformation as a result of AIoT's ability to propel proactive healthcare tactics that go beyond reactive interventions and into the domain of early identification and prevention. The insights revealed inside the AIoT ecosystem are poised to affect the future of healthcare in ways that were previously unthinkable as technology advances and AI algorithms become more complex.

Use Case: Data Collection and Integration in AIoT Healthcare: Unveiling Insights for Diagnosis and Treatment

The convergence of AI and the IoT has given rise to a paradigm-shifting innovation—the AIoT—in the field of contemporary healthcare. Predictive analytics becomes a potent tool inside this dynamic environment, revolutionising how healthcare is handled and delivered. Healthcare practitioners can detect possible health risks and take early action by using AIoT-driven predictive analytics, which is powered by real-time sensor data, previous patient data, and cutting-edge machine learning algorithms. A more robust and effective healthcare system might be developed as a result of this change from reactive to proactive treatment, which has the potential to transform patient outcomes and save healthcare costs.

The merge of AI and the IoT has ushered in a new era of data-driven patient care in the dynamic and always-changing world of contemporary healthcare. The seamless integration of data obtained from a complex network of devices and sensors inside the AIoT framework is at the core of this paradigm shift. The gathering, compilation, and analysis of health-related data have undergone a seismic shift as a result of this convergence, paving the way for sophisticated, highly individualised, and profoundly effective diagnosis and treatment approaches. Healthcare professionals will be able to access a wealth of insights that were previously concealed inside the large sea of data by utilising the interconnection of IoT and the analytical capability of AI.

8.3.5.1 The IoT ecosystem is unveiled in "a symphony of data streams"

The IoT ecosystem is a vast and complex network of linked sensors and devices with a wide range of uses, including smart homes and industrial automation. This ecosystem has a special significance in the context of healthcare. A dynamic tapestry of health-related data is produced by wearable technology, medical equipment, EHRs, home monitoring systems, and even the world of social media platforms. This tapestry encompasses a wide range of indicators, including specific lifestyle behaviours, medication adherence patterns, patient-reported results, and physiological measures like heart rate, blood pressure, and body temperature.

8.3.5.2 The alchemical transformation: gathering data from various sources

When these ostensibly unrelated data sources are expertly stitched into a coherent fabric, the actual power of AIoT in healthcare is revealed. In the past, patient health data was frequently fragmented and kept in separate silos by multiple healthcare organisations and systems. These silos are broken down by the AIoT ecosystem, ushering in a time when data streams from wearable technology integrate with past patient information, laboratory results, and lifestyle details. This interconnected mosaic of data gives medical practitioners a comprehensive insight into patients' particular health trajectories, enabling them to make wise decisions based on this knowledge.

8.3.5.3 AI-infused alchemy: transforming raw data into useful insights

However, raw data only has value if it is converted into usable insights, regardless of how plentiful or diverse it may be. Here, the joint symphony of AI and IoT really rings true. AI-powered algorithms are capable of processing and analysing the vast amounts of data produced by the AIoT ecosystem, extracting complex patterns, correlations, and insights that may be difficult for humans to understand. Deep learning models, neural networks, and machine learning algorithms become the maestros capable of spotting patterns that ultimately point to important conclusions by unravelling complex linkages within the data. Imagine an environment where patient-reported outcomes, EHR data, and wearable device data all come together for AI-driven analysis. The algorithms identify a complex relationship between the level of physical activity a patient engages in and the course of a chronic disease. This knowledge enables medical practitioners to design personalised therapy programmes that emphasise physical exercise, possibly reversing the trajectory of the condition's advancement. Such findings transform therapeutic procedures, transcending generalised methods and adjusting interventions to the unique narrative of each patient.

8.3.5.4 Transcending reactive healthcare: embracing proactive paradigms

The groundwork for a fundamental change towards proactive healthcare is laid out by the insights generated by AI-driven analysis of compiled health data. Healthcare treatments have typically taken the form of reactive measures, attending to symptoms or crises as they develop. Healthcare professionals can predict potential health risks before they develop overtly because of the AIoT, which is powered by data-driven insights. Practitioners can intervene early, avoiding problems and improving patient outcomes, by spotting deviations from normative patterns and associating them with particular health issues.

Imagine a situation where a patient is continuously monitored and is at risk for a cardiac event due to factors like age, family history, and lifestyle choices. Subtle changes in the patient's heart rate variability that could elude conventional examinations are revealed by AI analysis. These shifts are seen by the AI algorithms as indicators of oncoming cardiac discomfort, which prompts healthcare professionals to start actions to prevent the event from happening. By avoiding urgent procedures, this proactive healthcare strategy not only prolongs lives but also reduces healthcare costs.

8.3.5.5 Data sanctity and ethical beacon

While AIoT-powered healthcare is brilliant, it's crucial to negotiate ethical dilemmas and the data privacy landscape that comes with the collection and analysis of private patient information. It is essential to guarantee patient consent, protect data security, and follow strict privacy laws. For AIoT to be seamlessly woven into the fabric of healthcare, there must be a balance between potential benefits and the protection of individual rights.

8.3.5.6 Discovering the power of uncovered insights

An era of precise healthcare diagnostics and treatment has begun thanks to the marriage of AI-powered analysis and health data produced by the IoT network. The IoT ecosystem's intrinsic interconnection makes it possible to combine various data sources, creating a comprehensive depiction of patients' health profiles. This collection of data transcends its basic form through the lens of AI-driven algorithms, metamorphosing into useful insights, exposing connections, exposing trends, and articulating patterns that herald a new era of diagnosis and treatment choices. The ability of the AIoT to support preventative healthcare measures sets the stage for a transformational shift in patient care from responsiveness to preemption. The developments within the AIoT ecosystem are destined to shape the future of healthcare, giving rise to possibilities that were previously only imagined as technology advances and AI algorithms become more sophisticated.

Examples of Input

- The data for each patient contains a number of cumulative heart rate and temperature readings.

Output Diagrams

1. Bar graph showing the most recent heart rate readings
   Figure 8.5 displays the most recent heart rate readings for various patients. The x-axis represents each patient, and the y-axis shows the patient's most recent heart rate value. The various bars allow for a rapid comparison of the patients' present heart rates. This graph makes it easier to spot differences in the patients' heart rates.
2. Temperature Trends Line Plot
   Figure 8.6 shows how the temperatures of various patients have changed over time. The graph shows the temperature readings for

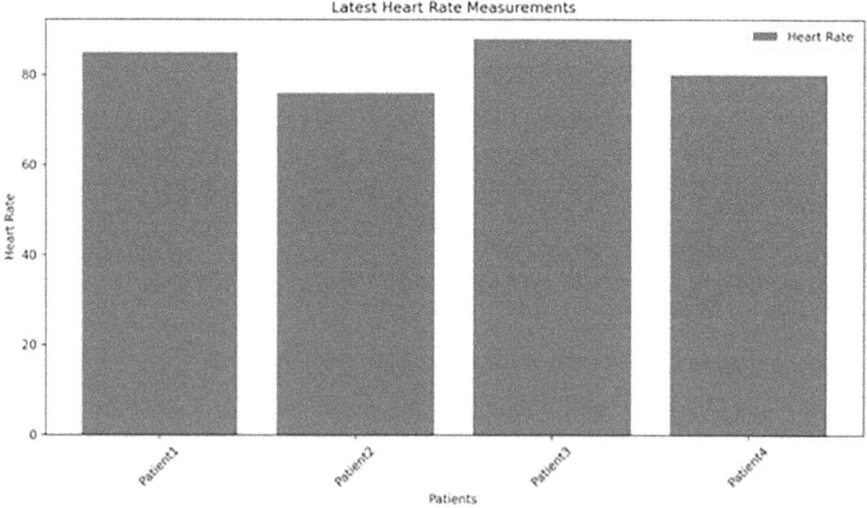

Figure 8.5 Heart rate comparison based on random data of patients.

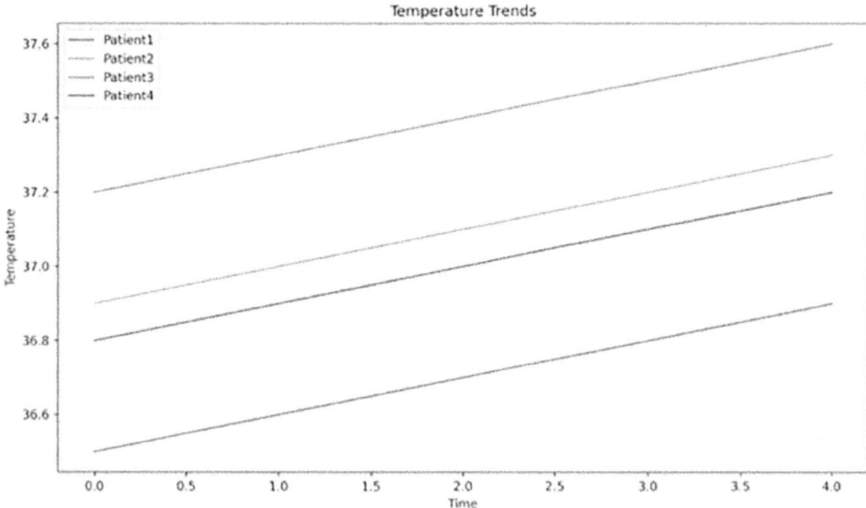

Figure 8.6 Temperature trend of the patients.

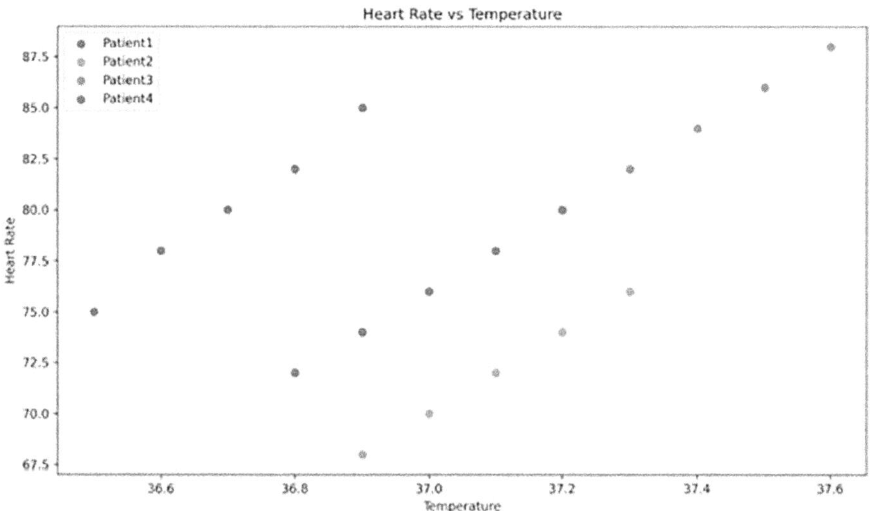

Figure 8.7 Heart rate versus temperature scatter plot of the sample size patients.

each patient as a line. Time is represented on the x-axis, and temperature values are represented on the y-axis. The trajectories of the lines reveal the temperature variations for each patient. This graph makes it possible to monitor temperature trends and spot any unusual deviations.

3. Heart Rate versus Temperature Scatter Plot

Figure 8.7 illustrates how different patients' temperatures and heart rates relate to one another. The data for each patient is shown as a set of points on a graph, where the temperature is represented by the *x*-coordinate and the heart rate by the *y*-coordinate. The scatter plot sheds light on possible relationships between temperature and heart rate. Clusters of data from various patients may show themes or patterns that could be investigated further.

8.4 PREDICTIVE ANALYTICS AND DIAGNOSTICS IN AIoT HEALTHCARE: PIONEERING PROACTIVE PATIENT CARE THROUGH DATA-DRIVEN INSIGHTS

Patient treatment has typically been reactive in the traditional healthcare model, with interventions taking place after symptoms or consequences have already been shown. Predictive analytics in the AIoT ecosystem, however, offers up a new frontier, allowing medical professionals to forecast future health risks before they manifest clinically [18]. This change is supported by the incorporation of patient history data as well as real-time sensor data from wearable technology and medical equipment. Predictive analytics makes healthcare a proactive endeavour by employing cutting-edge machine learning algorithms to analyse this plethora of data.

Figure 8.8 illustrates the idea of diagnostics and predictive analytics in AIoT healthcare. It exemplifies the combination of real-time sensor data, previous patient data, and sophisticated machine learning algorithms. The illustration shows how predictive analytics can foresee impending health problems, allowing medical professionals to act quickly and improve patient outcomes.

False positives or negatives in diagnostics and predictive analytics in AIoT healthcare put patient care at risk. Clinical validation studies, openness, and ongoing validation are all necessary to guarantee reliability. Models are improved by feedback loops, and risks are reduced by ensemble learning. Accuracy and compliance are maintained through strong data governance, human control, and quality assurance. Reliability is further improved by interdisciplinary cooperation and algorithmic bias prevention. Through the integration of these measurements, the reliability of AI-driven predictions

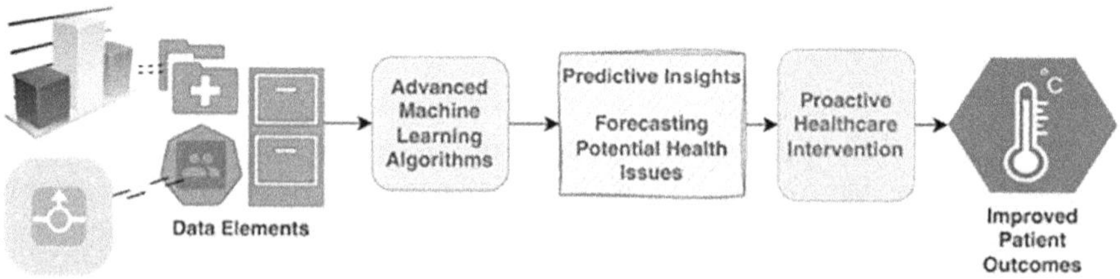

Figure 8.8 Predictive analytics and diagnostics in AIoT healthcare.

is increased, leading to proactive patient care through data-driven insights and reducing the possible negative effects of inaccurate forecasts.

8.4.1 The power of predictive analytics: transforming healthcare paradigms

A wealth of knowledge exists in historical patient data that can be used to evaluate health trajectories, illness progression, and therapeutic response. The integration of this historical data with AIoT-driven predictive analytics allows healthcare professionals to spot patterns, trends, and correlations that could otherwise be missed. Practitioners are better equipped to customise therapies based on unique risk profiles thanks to this comprehensive perspective of patients' medical histories [19]. Take a patient with a history of respiratory problems, for instance. AI-driven algorithms are able to recognise recurrent patterns that indicate an imminent respiratory episode by examining historical data pertaining to prior exacerbations, medication adherence, and environmental factors [20]. Armed with this knowledge, healthcare professionals can develop preventative measures that lower the risk, such as modifying drug dosages, offering tailored lifestyle advice, or suggesting environmental changes. Also an automatic disease detection system can help doctors to identify diseases more quickly and accurately, which can lead to earlier treatment and a lower death rate [21,22].

8.4.2 The patient's health through real-time sensor data

Predictive analytics gain a dynamic dimension from the real-time sensor data gathered from wearable tech and medical equipment. Such factors as heart rate, blood pressure, glucose levels, and even walking patterns are regularly monitored by these devices. Healthcare professionals may monitor their patients' physiological reactions to numerous stimuli and activities thanks to the seamless integration of this real-time data by the AIoT ecosystem. Think about a diabetic patient who wears a glucose monitor [23]. An AI-powered platform receives real-time glucose level data from this gadget. The AI algorithms may foresee fluctuations that might signify hyperglycaemic or hypoglycaemic episodes by comparing the present glucose levels with past data as well as contextual elements like meal consumption and physical activity [24]. To avoid harmful glucose spikes or crashes, healthcare professionals can then act with prompt advice or adjustments to insulin dosages.

8.4.3 Early identification and proactive measures

The ability to spot early indications of aberrations from regular physiological rhythms is the defining feature of AIoT-driven predictive analytics. These changes, which are frequently imperceptible and challenging to spot using conventional techniques, can signal imminent health problems. Predictive analytics recognises these variations and notifies healthcare professionals of potential hazards by utilising machine learning algorithms that are excellent at pattern identification. Consider a patient who has cardiovascular risk factors and wears a smartwatch that can continuously monitor their EKG. The AI algorithms examine the heart rate data and look for anomalous patterns that could indicate a higher risk of cardiac arrhythmias. When medical professionals spot this early indication, they can start actions including modifying drug dosages, suggesting lifestyle modifications, or scheduling more diagnostic tests. This prompt reaction may be able to stop major cardiac events and enhance patient outcomes.

8.4.4 The ripple effect's impact on healthcare costs and efficiency

Healthcare efficiency and costs are affected in a cascading manner by the shift from reactive to proactive care, which is driven by AIoT-based predictive analytics. Healthcare professionals can cut down on the number of ER visits, hospital stays, and invasive procedures by spotting possible health problems before they become serious. In addition to improving patient outcomes, this lessens the burden on healthcare expenditures and resources.

The transition to predictive therapies also makes it possible for healthcare organisations to distribute resources more effectively. Based on predictive models that anticipate patient needs, hospitals may optimise bed utilisation, staffing, and equipment upkeep. By reducing waste and ensuring that resources are directed where they are most needed, this proactive strategy improves the overall efficiency of healthcare systems.

8.4.5 Challenges and considerations

While AIoT-driven predictive analytics have enormous potential for the healthcare industry, it's critical to negotiate issues like data quality, algorithm accuracy, and data privacy. The accuracy of predictive models depends on the data they are trained on. For forecasts to be valid, past patient data must be verified for correctness, dependability, and representativeness. Additionally, as algorithmic procedures and findings must be transparent and interpretable to maintain patient and healthcare professional trust, predictions have a significant impact on therapeutic decisions. Making forecasts regarding people's health statuses is especially important when ethical considerations are involved. It is difficult to strike a balance between the potential advantages of predictive analytics and the dangers of unexpected effects or incorrect diagnosis. In the AIoT healthcare environment, it is crucial to protect patient privacy, get informed permission, and provide equal access to predictive therapies.

8.4.6 A proactive healthcare future

A future of proactive and individualised patient care is being ushered in by the integration of real-time sensor data and AI-driven predictive analytics within the AIoT healthcare framework. With this transformation, healthcare moves from being reactive to being proactive in identifying and resolving future health problems. Healthcare workers can respond early, minimising risks, improving patient outcomes, and lowering healthcare costs by utilising previous patient data and real-time insights. Proactive interventions have a wider impact than just on a single patient; they optimise resource allocation and boost the effectiveness of healthcare systems as a whole. Predictive analytics shines as a ray of hope as the AIoT ecosystem develops. It is a tool that enables healthcare professionals to embrace a new era of healthcare that is characterised by foresight, resilience, and improved patient well-being.

> **Use Case**: Predictive Analytics and Diagnostics in AIoT Healthcare: Pioneering Proactive Patient Care through Data-Driven Insights
>
> The IoT and AI have come together to create the AIoT, a paradigm that is full of innovation in the quickly changing world of modern healthcare. The idea of predictive analytics emerges as a beacon of transformation inside this dynamic ecosystem, revolutionising the very foundation of healthcare delivery. Healthcare providers may identify and prevent possible health risks thanks to AIoT-driven predictive analytics, which enables a change from reactive to proactive patient care by utilising real-time sensor data, previous patient data, and cutting-edge machine learning algorithms. This change has the potential to redefine patient outcomes, control healthcare spending, and create a more robust and effective healthcare system.

8.4.6.1 Predictive analytics: a healthcare evolution catalyst

Traditional healthcare models frequently take a reactive stance, in which interventions are initiated after symptoms or consequences show. Predictive analytics inside the AIoT ecosystem, however, moves healthcare into an anticipatory space where practitioners can foresee health issues before they worsen thanks to data-driven insights. The seamless fusion of historical patient data and real-time sensor information from wearable technology and medical equipment is what is driving this paradigm change. Predictive analytics turns healthcare from a passive response model into an area of proactive intervention through the application of cutting-edge machine learning algorithms.

8.4.6.2 Unveiling the power of historical patient data

Historical patient data is a gold mine of information that can be used to understand disease progression, health trajectory changes, and therapy response. By integrating this historical data, AIoT-driven predictive analytics enables healthcare practitioners to identify patterns, trends, and correlations that can evade human observation. Healthcare professionals can now personalise therapies based on unique risk profiles thanks to this comprehensive perspective of patients' medical histories.

Imagine a situation where a patient who has a history of respiratory problems is involved. AI algorithms reveal recurring patterns that indicate an oncoming respiratory episode by analysing historical data spanning past exacerbations, medication adherence patterns, and the impact of environmental factors. Armed with this knowledge, healthcare professionals can devise preventative measures that lower the risk, such as adjusting medication dosages, providing tailored lifestyle advice, or recommending changes to the patient's surroundings.

8.4.6.3 Real-time sensor insights: windows into the dynamics of health

Real-time sensor inputs from wearable technology and medical equipment provide another level of dynamism to the field of predictive analytics. Critical factors including heart rate, blood pressure, glucose levels, and even walking patterns are routinely monitored by these devices. Healthcare practitioners now have a real-time window into patients' physiological reactions to a variety of stimuli and activities thanks to the seamless integration of this real-time data by the AIoT network.

Imagine a patient fighting diabetes and equipped with a wearable glucose meter. An AI-powered platform receives real-time glucose level data from this gadget. AI systems can predict swings that could indicate oncoming hyperglycaemic or hypoglycaemic episodes by comparing the present glucose levels with past data and contextual factors, such as meal consumption and physical activity. Healthcare professionals can then intervene with prompt advice or adjustments to insulin levels, preventing harmful glucose rises or crashes.

8.4.6.4 Early detection and proactive measures are at the forefront

Predictive analytics powered by AIoT is renowned for its skill in identifying early departures from typical physiological trends. These aberrations, which are frequently subtle and evade typical assessment techniques, can act as early warning signs of possible health issues. Predictive analytics detects these aberrations and notifies healthcare practitioners of potential hazards by utilising machine learning algorithms skilled at pattern identification.

Imagine a patient who has cardiovascular risk factors and is equipped with a smartwatch that can continuously monitor their EKG. AI algorithms analyse heart rate data to find abnormal patterns that could indicate increased risks for cardiac arrhythmias. Healthcare practitioners can start measures, such as medication modifications, lifestyle change advice, or extra diagnostic testing, once they are aware of these early symptoms. This quick reaction may be able to prevent serious cardiac events, greatly improving patient outcomes.

8.4.6.5 *Increasing efficiency and reducing costs: the ripple effect*

Predictive analytics powered by AIoT enable the shift from reactive to proactive care, which has an impact on healthcare efficiency and costs. Healthcare professionals can cut down on the number of ER visits, hospital stays, and invasive procedures by identifying potential health issues in the early stages. This transformation improves patient outcomes while easing the burden on healthcare finances and resources.

The paradigm shift towards predictive interventions also enables healthcare organisations to more precisely manage resources. Based on predictive models that forecast patient needs, hospitals can optimise bed utilisation, personnel arrangements, and equipment maintenance. By reducing waste and ensuring that resources are used where they are most needed, this proactive approach increases the overall effectiveness of healthcare systems. Despite the enormous promise of AIoT-driven predictive analytics in healthcare, it is critical to overcome issues with data quality, algorithmic accuracy, and data privacy. The calibre of the data that predictive models are trained on has a profound impact on their effectiveness. For forecasts to be accurate, past patient data must be maintained with fidelity, consistency, and relevance. Additionally, algorithm transparency is important since predictions affect healthcare decisions. Ethics issues are crucial, particularly when predicting people's health conditions. It is a delicate dance to strike a balance between the benefits of predictive analytics and the risk of unexpected effects or incorrect diagnosis. Within the AIoT healthcare space, it is crucial to protect patient privacy, obtain informed permission, and provide equal access to predictive interventions.

The integration of real-time sensor data and AI-driven predictive analytics inside the AIoT healthcare framework propels us into a future characterised by proactive and individualised patient care. With this shift, healthcare moves from being a reactive industry to one that anticipates and averts looming health problems. Healthcare professionals may quickly intervene, reducing risks, improving patient outcomes, and lowering healthcare costs by leveraging previous patient insights and real-time disclosures. Proactive interventions have far-reaching effects that go beyond specific patients, optimising resource allocation and boosting the general effectiveness of healthcare systems. Predictive analytics is a technology that enables healthcare practitioners to embrace an era where foresight, resiliency, and patient well-being dictate the trajectory of healthcare's transformation as the AIoT ecosystem advances.

Sample Output:
>Predictive Analysis Alerts:
>Patient 1: Fever detected
>Patient 2: Fever detected
>Patient 3: Fever detected
>Patient 4: Fever detected
>Patient 5: Fever detected
>Patient 6: Fever detected
>Patient 7: Fever detected
>Patient 8: Fever detected
>Patient 9: Fever detected
>Patient 10: Fever detected

Figure 8.9 shows each patient's temperature trend is shown by a separate line. The lines provide a visual picture of the variability in the patients' temperature patterns by displaying how their temperatures change over time.

The predictive analysis carried out by the programme relies heavily on the graph. The code sample lays a framework for identifying increased temperatures that may suggest fever in patients, even though it doesn't explicitly present alerts based on temperature trends. The programme would send

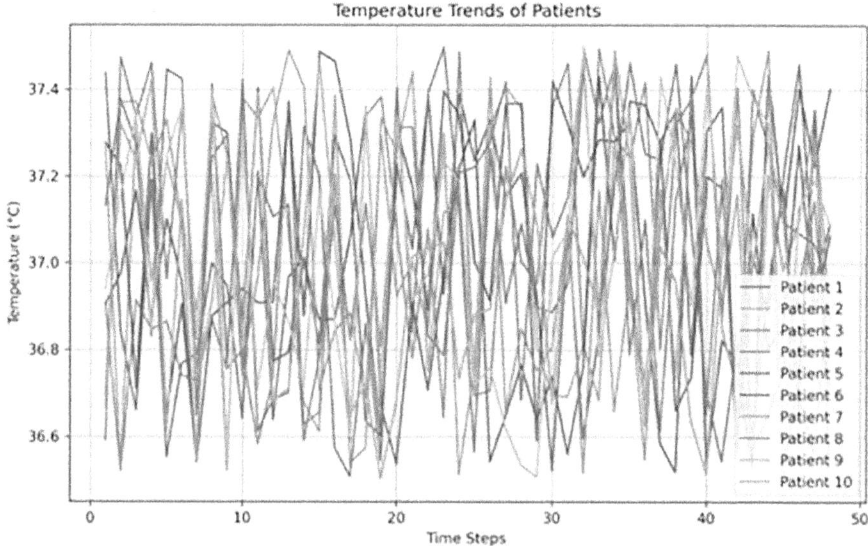

Figure 8.9 Temperature trend of each patient on the sample size.

out an alert announcing the discovery of a fever if a patient's temperature exceeded a predetermined threshold (37.2°C in this example).

Healthcare workers can rapidly evaluate and compare the temperature changes of various patients thanks to the graph. It offers a simple visualisation that makes it easier to spot trends, patterns, or anomalies that could point to underlying medical issues.

8.5 AIoT-ENABLED DISEASE MANAGEMENT: REVOLUTIONISING CHRONIC CARE THROUGH PERSONALISED INSIGHTS

The management and treatment of chronic diseases have been redefined as a result of the fusion of AI and the IoT [25]. Chronic diseases, which include illnesses like diabetes, hypertension, and respiratory ailments, place a heavy load on the world's healthcare systems. A paradigm change in the treatment of chronic diseases has been made possible by the introduction of AIoT, enabling patients to actively participate in their own health. AIoT-enabled illness management has the possibility of revolutionising patient care, increasing treatment outcomes, and reducing the burden on healthcare systems by utilising smart devices, continuous health monitoring, and AI-driven insights.

Figure 8.10 highlights how patients with chronic diseases use smart devices for health tracking and personalized advice, demonstrating AIoT-enabled disease management. Figure 8.10 shows a patient interacting with wearable technology, smart medical equipment, and IoT-enabled tools to track health indicators and follow treatment regimens.

Strategies for patient participation are essential for the successful deployment of AIoT devices in illness management. These include real-time feedback, social support networks, gamification for motivation, and individualised instruction. The continuity of care is guaranteed via integration with healthcare providers, and usability is improved through user-centric design. Concerns are addressed with follow-up and ongoing support. These tactics maximise the potential of AIoT devices in disease management by empowering and motivating patients to actively participate in their care. This improves adherence, improves health outcomes, and increases patient satisfaction with care.

8.5.1 The chronic disease challenge: a call for innovation

Globally, chronic diseases are the main cause of morbidity, mortality, and healthcare spending. They are frequently characterised by their protracted duration and slow advancement. Complex and diverse strategies, such as

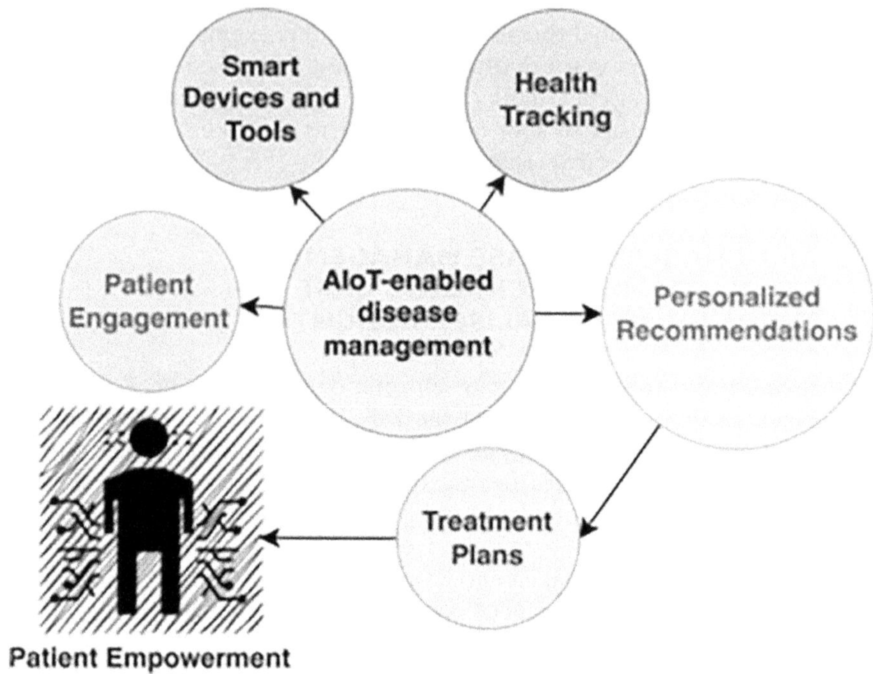

Figure 8.10 AIoT-enabled disease management.

medication adherence, lifestyle changes, and routine monitoring, are used to manage chronic illnesses. Patients used to only occasionally communicate with healthcare professionals, frequently only when complications or symptoms increased [26]. However, the environment is changing as AIoT technologies introduce a new aspect of proactive, personalised, and data-driven chronic disease management.

8.5.2 Empowering patients through smart devices

The use of smart gadgets that smoothly integrate into patients' lives is at the heart of AIoT-enabled disease management. Individuals are given the ability to continuously monitor their health indicators and vital signs thanks to these gadgets, which range from wearable sensors to connected medical equipment. These gadgets turn into indispensable aids for individuals with chronic illnesses like diabetes or hypertension that help them fill the time in between healthcare sessions. Think about a diabetic patient who wears a continuous glucose monitoring (CGM) system, a wearable gadget that monitors blood sugar levels continuously throughout the day [27]. A smartphone app receives real-time data from the CGM and uses AI algorithms to examine trends and swings. Patients receive knowledge about their glucose patterns, empowering them to choose food, exercise, and medicine wisely. Through a continuous feedback loop, disease management is transformed from a reactive to a proactive strategy, enabling patients to take action early and avoid significant changes in their health.

8.5.3 Monitoring continuously to ensure prompt action

The ability to continuously monitor is one of the most important benefits of AIoT-enabled illness management. Subtle changes in health indicators that are frequently associated with chronic diseases can go unreported until issues occur. Patients may continuously monitor their vital signs, medication compliance, and lifestyle choices thanks to AIoT. When this data is combined, AI algorithms look for variations from the usual, which results in alarms being sent to patients and healthcare professionals. Using a smart blood pressure monitor that wirelessly communicates readings to a central platform, for instance, a patient with hypertension might use it. Both the patient and the healthcare professional are alerted if the AI algorithms find persistent blood pressure spikes that signify inadequate disease control. This prompt action could stop the progression of problems from hypertension, like heart disease or stroke. AI-driven insights and ongoing monitoring enable patients to take preventative action while providing healthcare providers with real-time information for data-driven decision-making.

8.5.4 Personalised treatment plans: a shift from one-size-fits-all

An era of individualised treatment programmes catered to the needs of specific patients may soon be here thanks to AIoT-enabled illness management. In the past, treatment plans for chronic illnesses were frequently generalised without taking into consideration the particular traits and reactions of each patient. Precision medical techniques are made possible by the incorporation of AIoT technology, which allows for a more detailed understanding of patient health profiles. Imagine an asthma patient utilising a sensor-equipped linked inhaler. This device monitors symptom occurrences, environmental triggers, and inhaler usage trends. The data is analysed by AI systems, which find relationships between particular triggers and exacerbations. Patients receive personalised instructions to avoid triggers and modify medication dosages armed with this knowledge. This individualised approach not only promotes patients' quality of life but also enhances disease control.

8.5.5 Reducing healthcare burden and costs

Due to their long-term management, frequent hospitalisations, and related complications, chronic diseases place a major load on healthcare systems. By encouraging improved disease control and lowering the frequency of acute exacerbations, AIoT-enabled disease management has the potential to lessen this burden. The need for emergency medical interventions declines, which results in lower healthcare costs, when people actively manage their diseases and receive timely interventions based on AI-driven insights. Additionally, AIoT-enabled disease management's proactive approach can lessen the load on healthcare institutions. Hospitals can optimise resource allocation and lessen the stress on overworked healthcare systems by avoiding problems and hospitalisations. The cost savings brought about by AIoT-driven actions can be used to fund the advancement of medical research, better patient care, and access to disadvantaged groups.

8.5.6 Challenges

Despite the alluring promises of AIoT-enabled illness management, a number of obstacles need to be overcome for widespread use. Due to the digital transmission and storage of sensitive health information, data privacy, security, and informed permission are crucial. To keep people's faith in AIoT technologies, it's imperative to make sure that patient data is safeguarded against unauthorised access. Another issue is interoperability because patients might utilise various gadgets made by various companies. For thorough disease control, it is crucial to make sure that these devices can communicate and share data without any issues. This problem is being addressed, and attempts are being made to standardise the environment.

8.5.7 Empowering patients, transforming care

The ability of AIoT to facilitate disease management is evidence of how technology has the potential to transform healthcare. Patients with chronic conditions can take charge of their own health management by combining the capabilities of AI, IoT, and continuous health monitoring. The way chronic illnesses are approached, treated, and prevented is changing as a result of personalised insights, real-time monitoring, and proactive interventions. The AIoT ecosystem offers a beacon of hope as healthcare systems battle the difficulties brought on by chronic diseases. It provides a chance to build a future in which patients lead better, more empowered lives and in which the financial burden of chronic diseases on healthcare systems is greatly reduced.

8.6 TELEMEDICINE AND VIRTUAL HEALTH PLATFORMS IN THE AIoT ERA: REDEFINING HEALTHCARE ACCESS AND DELIVERY

A disruptive era in healthcare has begun with the introduction of AI in everything (AIoT), which has sparked the growth of telemedicine and virtual health platforms. These ground-breaking solutions are changing how people access healthcare, get medical advice, and interact with healthcare professionals remotely. Telemedicine and virtual health platforms are overcoming geographic obstacles, improving patient outcomes, and increasing the effectiveness of healthcare delivery by utilising the synergy of AI and the IoT [28]. By providing real-time data transfer, AI-driven insights, and individualised care experiences that change the healthcare landscape, this convergence empowers patients and healthcare professionals alike.

Reasonable Internet activities, digital literacy programmes, and mobile health solutions are some of the techniques being used to lessen the impact of the digital divide on access to AIoT-enhanced telemedicine. Public Wi-Fi locations can be set up and fair device access can be provided through community collaborations. Inclusion is guaranteed by outreach programmes and multilingual assistance. It is essential to advocate for laws that promote digital inclusiveness. By enabling marginalised groups to use telemedicine services, these policies improve healthcare outcomes and lessen inequities.

8.6.1 Healthcare beyond borders: the telemedicine revolution

The delivery of healthcare services remotely is known as telemedicine, and in recent years it has expanded at an unprecedented rate. The AIoT has been instrumental in this rise. Giving patients the option to get medical care and consultations without having to be physically present is the fundamental tenet of telemedicine. People who live in distant locations, have mobility issues, or have trouble accessing standard healthcare services would especially benefit from this. Patients can communicate with healthcare professionals via video conferencing, audio conferences, and messaging platforms thanks to AIoT-enabled telemedicine. The incorporation of IoT-enabled devices that communicate real-time health data, such as vital signs, glucose levels, and medication adherence, enhances these interactions. This data flow not only makes it possible to comprehend patients' health conditions more thoroughly, but it also provides the basis for AI-driven diagnosis and therapy choices [29].

8.6.2 Enhancing virtual consultations with real-time data transmission

Telemedicine consultations are more effective and of higher quality when IoT-enabled devices are integrated. Patients who have wearable sensors, smart medical equipment, and monitoring tools at their disposal can instantly submit information to healthcare professionals. Healthcare practitioners can make judgements without being physically close to patients thanks to this real-time data transfer, which gives them a dynamic snapshot of their health states. Imagine a patient receiving a telemedicine consultation for a persistent respiratory ailment. Real-time breathing data from the patient is sent to the interface of the healthcare professional via an IoT-enabled spirometer, a device that monitors lung function. The data is analysed by AI algorithms, which identify trends in lung function, possible exacerbations, and treatment outcomes. The healthcare professional can modify drugs, suggest interventions, and personalise treatment plans in light of these insights without requiring an in-person visit.

8.6.3 Enhancing virtual consultations with AI-driven diagnostics

The combination of AI and telemedicine adds a revolutionary dimension to virtual consultations—an improvement in diagnostic precision and accuracy. AI algorithms use powerful machine learning techniques to analyse the real-time health data that IoT-enabled devices transmit. These algorithms find patterns, correlations, and abnormalities that conventional assessment techniques might not be able to spot. Think of a situation when a patient receiving telemedicine consultation has cardiovascular risk factors. The patient's heart rate, blood pressure, and physical activity are monitored using IoT-enabled wearable technology. AI systems examine this data and find minute variations in heart rate that are related to cardiac health. The healthcare provider can then start additional investigations or interventions if the algorithms notice variations from typical trends. The diagnostic layer powered by AI improves the effectiveness of virtual consultations by encouraging quicker and more precise diagnosis.

8.6.4 Plans for individualised treatment: adapted through virtual interactions

Healthcare professionals are now better equipped to deliver individualised treatment regimens through virtual encounters thanks to the convergence of telemedicine and AIoT. Historically, therapeutic strategies were frequently generalised, ignoring the particular requirements and reactions of different patients. However, telemedicine supported by the AIoT brings a lot of data that supports precision medicine methods. For instance, a diabetic patient participating in a telemedicine session can send real-time blood sugar levels using IoT-enabled glucose monitors. Along with history information, meal records, and lifestyle habits, AI systems analyse these readings. The healthcare professional can create a customised treatment plan using this comprehensive picture, taking into account the patient's particular dietary preferences, medication responses, and glucose swings. The end result is a treatment plan that maximises health outcomes while being in line with the patient's preferences.

8.6.5 Bridging the gap between healthcare equity and accessibility

The ability of AIoT-enabled telemedicine to close healthcare access gaps and advance equality is one of the technology's most spectacular effects. Healthcare disparities are frequently caused by geographic restrictions, inadequate transportation, and a lack of medical facilities, especially in rural and underdeveloped areas. Patients can obtain specialised treatment and expertise without leaving their communities thanks to telemedicine, which is supported by the AIoT ecosystem. Additionally, telemedicine makes healthcare more easy and accessible by reducing the amount of time patients must miss from work or travel for appointments. The elderly and people with disabilities, among other vulnerable populations, stand to gain a lot from this improved access to care.

While telemedicine provided by the AIoT has a lot of potential, there are a few issues that need to be addressed. Patients' capacity to use IoT devices and participate in virtual consultations efficiently may be impacted by their level of digital literacy, particularly among elderly populations. The key to overcoming this barrier is to provide user-friendly interfaces and provide instructions on how to operate the device. The privacy and security of data are of utmost importance when sending private medical data over the Internet. To protect patient data, it is crucial to set up reliable encryption techniques, get informed consent, and follow privacy laws.

The delivery of healthcare, accessibility, and patient empowerment are being redefined through telemedicine and virtual health platforms, driven by the AIoT. Transcending geographical borders, improving diagnosis accuracy, and advancing healthcare equity are all made possible by the integration of real-time data transfer, AI-driven diagnostics, and personalised treatment plans within virtual interactions. By enabling patients to interact with healthcare providers in novel and disruptive ways, the AIoT ecosystem has democratised healthcare. A future in which healthcare is not limited to physical locations but instead is available to patients whenever they need it and wherever they are, thanks to the synergy of AIoT and telemedicine, has the potential to be created as this fusion continues to develop.

8.7 ETHICAL AND PRIVACY CONSIDERATIONS IN AIoT HEALTHCARE: NAVIGATING THE COMPLEX TERRAIN OF DATA SECURITY AND PATIENT PRIVACY

Unquestionably, AIoT is revolutionising patient care, diagnosis, and treatment in the healthcare industry. But although the growth of AIoT devices produces a plethora of health-related data, it also opens up a whole new world of ethical questions and privacy concerns [30]. To balance using the power of data-driven healthcare while protecting sensitive patient information, the seamless integration of AI and the IoT provides a complex terrain that requires cautious navigation [31]. In the middle of this digital transformation, patient privacy, strong data security, ethical frameworks, and informed consent assume a central role as essential elements for enabling the responsible and successful integration of AIoT technologies in healthcare systems.

For ethical AIoT implementation in healthcare, established frameworks or tailored guidelines are imperative. These should prioritise patient well-being, respect autonomy, ensure fairness, and safeguard privacy. Transparency, accountability, and continuous ethical review are essential. Benefit-risk assessments and stakeholder engagement aid in decision-making. Education and training foster ethical awareness. Upholding these principles ensures that AIoT technologies are deployed responsibly, benefiting individuals and society while mitigating potential harms.

8.7.1 Ethics complexities revealed by the data deluge

Healthcare systems obtain previously unattainable insights into the health profiles of patients as a result of the flood of health-related data generated by AIoT devices, including wearable sensors and medical equipment. These discoveries could revolutionise patient treatment, but they also raise a number of ethical issues [32]. The level of detail in the data gathered prompts inquiries about ownership of the data, consent, and possible information misuse. In the AIoT healthcare ecosystem, finding a balance between data use and individual rights becomes of utmost importance.

8.7.2 Patient empowerment through informed consent

Informed consent is more crucial in the setting of AIoT healthcare. Patients must be informed completely and clearly about the sorts of data gathered, how they will be used, and any possible advantages and disadvantages. Patients who have given their informed consent are better equipped to decide whether or not to take part in initiatives for data gathering and exchange. Furthermore, consent includes the freedom to leave data-sharing agreements at any time and goes beyond simple agreement. One of the pillars of moral AIoT healthcare is making sure that patients are well-informed and have agency over how their data is used.

8.7.3 Data security: protecting private information

The enormous volumes of health-related data that AIoT devices produce are valuable to bad actors looking to exploit vulnerabilities as well as to healthcare providers. Data breaches may compromise patient privacy and put people at risk of identity theft or insurance fraud, among other serious repercussions. Sensitive health information must be protected from unauthorised access, breaches, and cyberattacks with strong data security procedures. The AIoT healthcare infrastructure must include encryption, secure communication methods, and strict authentication procedures.

8.7.4 Protecting patient privacy and confidentiality

In AIoT healthcare, protecting patient privacy is a moral requirement. There is an inherent risk of identifying specific persons based on patterns and correlations as AI systems process and analyse health data. To protect patient privacy, aggregated data must be anonymised and cleansed of personally identifiable information. Differential privacy is one example of a privacy-preserving technology that permits significant data analysis without re-identifying any particular people. Maintaining trust and ethical integrity depends on striking a balance between the usefulness of data and patient anonymity.

8.7.5 Algorithmic fairness and bias: addressing disparities

The AIoT healthcare algorithms are not immune to potential biases that may exist in historical data. For marginalised or underrepresented populations, these biases may lead to unjust predictions, wrong diagnoses, or unequal treatment. Identification and mitigation of biases, the development of varied and representative training datasets, and ongoing algorithm performance monitoring are all necessary to ensure algorithmic fairness. It is a requirement of ethical AIoT healthcare that technological innovations contribute to equitable and inclusive healthcare solutions rather than escalating already existing healthcare inequities.

8.7.6 Transparency and accountability: ensuring responsible AI

The opaque nature of AI systems prompts questions about accountability and transparency. Patients and healthcare professionals need to understand how data is processed and how AI-driven choices are made. Explainable AI, which sheds light on algorithmic judgement, can increase openness and foster patient confidence in AIoT healthcare systems. Furthermore, it's important to establish clear lines of accountability so that decisions made by AI algorithms can be attributed to the appropriate persons.

Regulatory frameworks and ethical considerations in AIoT healthcare are interwoven. Governments and healthcare organisations are crucial in setting the rules, regulations, and legislation that control data usage, patient privacy, and the use of ethical AI. By guaranteeing that patient rights are protected and data-driven innovations are in line with societal values, these frameworks offer a road map for moral AIoT healthcare practices.

8.7.7 The ethical imperative in AIoT healthcare

Integration of the AIoT in healthcare has the potential to significantly improve patient care, diagnosis, and treatment. There are still some privacy and ethical issues to be resolved in this shift, though. Given the abundance of health-related data that AIoT devices produce, ethical frameworks must govern data use, informed permission, and patient privacy protection. A careful balance between advancing technology and preserving individual rights is necessary for the ethical integration of AIoT technologies in healthcare. Healthcare systems may use the power of AIoT to usher in a new era of patient-centred, accountable, and data-driven healthcare that respects the rights and values of all people by giving ethical considerations top priority.

8.8 CHALLENGES AND FUTURE DIRECTIONS OF AIoT IN HEALTHCARE: NAVIGATING HURDLES AND PAVING THE PATH FORWARD

Patient care, diagnosis, and treatment have been redefined as a result of the integration of AIoT in healthcare. The adoption of AIoT in healthcare is not without its share of hurdles, despite the excitement and potential [33]. To enable the appropriate and successful integration of AIoT in healthcare systems, challenges such as interoperability, standardisation, regulatory compliance, data accuracy, and the rapid growth of technology must be overcome. Future directions of AIoT in healthcare hold the promise of more advanced AI models, improved data fusion techniques, and a cooperative synergy between technology developers and healthcare professionals. These developments will ultimately shape a healthcare landscape that is data-driven, effective, and patient-centred.

8.8.1 Standardisation and interoperability: the need for cohesion

Achieving interoperability and standardisation in the AIoT healthcare ecosystem is one of the biggest difficulties. A fragmented environment is created by the enormous variety of IoT devices, each of which has its own protocols, data formats, and communication techniques, making it difficult to interchange and integrate data in an efficient manner. To facilitate effective connection between many devices, common standards, communication protocols, and data formats must be established. This cohesiveness is essential to ensure that information from many sources may be combined, examined, and used in a coordinated manner to enable comprehensive patient care.

8.8.2 Data governance and regulatory compliance

The healthcare industry operates in a highly regulated environment that is governed by strict privacy and security laws including the General Data Protection Regulation and Health Insurance Portability and Accountability Act. Regulatory compliance is crucial as AIoT creates and processes enormous amounts of sensitive health data. Data governance, secure data storage, access controls, and ethical data usage must all be carefully considered to strike a balance between utilising the power of data-driven healthcare and protecting patient privacy.

8.8.3 Data reliability and accuracy

AIoT device data must be reliable and trustworthy to support precise diagnosis and well-informed decision-making. Any data inconsistencies, biases, or errors can provide inaccurate forecasts and put patient safety and healthcare results in danger. Strict sensor calibration, IoT device validation, and ongoing data quality monitoring are all required to ensure data accuracy. By finding and correcting inaccurate or inconsistent data points, the use of AI tools, such as anomaly detection algorithms, can improve data accuracy.

8.8.4 Considerations for social and ethical behaviour

Ethical and social issues become more important as AIoT technologies become more pervasive in healthcare. Careful thought must be given to issues such as data ownership, patient permission, equitable access, and the potential to exacerbate healthcare inequities. It is crucial to strike a balance between societal ideals and technology improvements to maximise intended effects and minimise unforeseen ones for AIoT-driven healthcare.

8.8.5 Future directions: an innovative road

Despite these obstacles, the application of AIoT in healthcare has promising future prospects. The development of AI models promises increasingly advanced predictive and diagnostic capabilities, driven by developments in machine learning and deep learning. These models will be able to identify tiny patterns, make complex predictions, and provide previously unfathomable insights. The integration of various data sources, including wearable technology, EHRs, social determinants of health, and genomic data, will be made possible through improved data fusion algorithms. By providing a thorough picture of patients' health profiles, this holistic approach would enable precision medicine and preemptive therapies catered to specific patient requirements. The future of AIoT in healthcare will be significantly shaped by collaboration between technology developers and healthcare experts. Interdisciplinary efforts will drive innovation, with technologists and medical professionals collaborating to create solutions that are in line with clinical realities, patient preferences, and healthcare workflows. AIoT technologies will be more than simply technologically cutting edge; they will also be useful and significant in actual healthcare settings thanks to this collaborative synergy [7,10,21,22,33,34].

The difficulties that come with the deployment of AIoT in healthcare are not insurmountable obstacles; rather, they present chances for development, creativity, and responsible change. The well-being of patients and data integrity must always come first when addressing interoperability, standardisation, data correctness, and ethical issues. Healthcare is set for a future characterised by data-driven insights, individualised care, and more effective healthcare delivery as AIoT technologies continue to advance. Healthcare institutions may embrace the potential of AIoT while preserving the ideals of ethics, privacy, and equitable access by overcoming these obstacles with effort and collaboration. The road ahead is one of opportunity and responsibility; it has the ability to transform healthcare for the benefit of people all around the world.

8.9 CONCLUSION

The convergence of AI and the AIoT has had a profound impact on the quickly developing field of healthcare. These developments open the way for proactive and individualised patient care in the future, revolutionising diagnostics and changing treatment paradigms. Predictive analytics, where historical patient data and real-time sensor insights combine, allows healthcare workers to anticipate probable health risks. This is made possible by the combination of AI and IoT. This ability to forecast outcomes signals a transition from reactive to proactive treatment, which will reduce costs, improve patient outcomes, and better utilise available resources. Additionally, continuous real-time health data collection is made possible by the AIoT fabric's combination of wearable technology and remote patient monitoring. Healthcare professionals are better able to manage resource restrictions and improve patient experiences thanks to remote monitoring.

AIoT interaction reveals a wealth of insights from intricate data streams. These insights enable knowledgeable judgements that cater to the needs of specific patients, from early detection to aggregated health data. The combination of AIoT and healthcare continues to hold endless possibilities as technology and AI algorithms develop. This change is not without difficulties, though. The importance of ethical considerations, data protection, and algorithm precision has not changed. For the seamless integration of AIoT technologies, it's critical to strike a balance between innovation and the protection of individual rights. The AIoT has the potential to completely change the healthcare sector. The AIoT is revolutionising how health is managed and patients are treated by utilising predictive analytics, remote monitoring, and data-driven insights.

REFERENCES

1. Ball, M.J. and Lillis, J., 2001. E-health: Transforming the physician/patient relationship. *International Journal of Medical Informatics*, 61(1), pp. 1–10.
2. Zhang, J. and Tao, D., 2020. Empowering things with intelligence: A survey of the progress, challenges, and opportunities in artificial intelligence of things. *IEEE Internet of Things Journal*, 8(10), pp. 7789–7817.
3. Letaief, K.B., Shi, Y., Lu, J. and Lu, J., 2021. Edge artificial intelligence for 6G: Vision, enabling technologies, and applications. *IEEE Journal on Selected Areas in Communications*, 40(1), pp. 5–36.
4. Sahu, M., Gupta, R., Ambasta, R.K. and Kumar, P., 2022. Artificial intelligence and machine learning in precision medicine: A paradigm shift in big data analysis. *Progress in Molecular Biology and Translational Science*, 190(1), pp. 57–100.
5. Pise, A.A., Almusaini, K.K., Ahanger, T.A., Farouk, A., Pareek, P.K. and Nuagah, S.J., 2022. Enabling artificial intelligence of things (AIoT) healthcare architectures and listing security issues. *Computational Intelligence and Neuroscience,* 1, p. 8421434.
6. Pathak, D., Rathod, V. and Chapaneri, R., 2023, June. A comprehensive survey on AIoT integrated smart health cities. In *AIP Conference Proceedings* (Vol. 2705, No. 1). AIP Publishing.
7. Pestana, M., Pereira, R. and Moro, S., 2020. Improving health care management in hospitals through a productivity dashboard. *Journal of Medical Systems*, 44(4), p. 87.
8. Cobiac, L.J., Magnus, A., Barendregt, J.J., Carter, R. and Vos, T., 2012. Improving the cost-effectiveness of cardiovascular disease prevention in Australia: A modelling study. *BMC Public Health*, 12(1), pp. 1–10.
9. Baker, S. and Xiang, W., 2023. Artificial intelligence of things for smarter healthcare: A survey of advancements, challenges, and opportunities. *IEEE Communications Surveys &Tutorials*, 25, p. 1261–1293.
10. H. D, A. D. S, A. Begum and P. Hemanth, 2023. Early detection of breast cancer with IoT: A promising solution. In *2023 International Conference on Advances in Computing, Communication and Applied Informatics (ACCAI)*, Chennai, India (pp. 1–6). https://doi.org/10.1109/ACCAI58221.2023.10199910.
11. Wu, C., Wang, B., Au, O.C. and Liu, K.R., 2022. Wi-fi can do more: toward ubiquitous wireless sensing. *IEEE Communications Standards Magazine*, 6(2), pp. 42–49.

12. Salih, K.O.M., Rashid, T.A., Radovanovic, D. and Bacanin, N., 2022. A comprehensive survey on the Internet of Things with the industrial marketplace. *Sensors*, 22(3), p. 730.

13. Lim, Y.J. and Kim, T.J., 2022. Digital healthcare development and mhealth in South Korea. In Kota Kodama, Shintaro Sengoku, *Mobile Health (mHealth) Rethinking Innovation Management to Harmonize AI and Social Design* (pp. 83–116). Singapore: Springer Nature Singapore.

14. Shi, Q., Zhang, Z., Yang, Y., Shan, X., Salam, B. and Lee, C., 2021. Artificial intelligence of things (AIoT) enabled floor monitoring system for smart home applications. *ACS Nano*, 15(11), pp. 18312–18326.

15. Qian, K., Zhang, Z., Yamamoto, Y. and Schuller, B.W., 2021. Artificial intelligence internet of things for the elderly: From assisted living to health-care monitoring. *IEEE Signal Processing Magazine*, 38(4), pp. 78–88.

16. Zhang, J. and Tao, D., 2020. Empowering things with intelligence: A survey of the progress, challenges, and opportunities in artificial intelligence of things. *IEEE Internet of Things Journal*, 8(10), pp. 7789–7817.

17. Zhang, Q., Jin, T., Cai, J., Xu, L., He, T., Wang, T., Tian, Y., Li, L., Peng, Y. and Lee, C., 2022. Wearable triboelectric sensors enabled gait analysis and waist motion capture for IoT-based smart healthcare applications. *Advanced Science*, 9(4), p. 2103694.

18. Akhigbe, B.I., Munir, K., Akinade, O., Akanbi, L. and Oyedele, L.O., 2021. IoT technologies for livestock management: A review of present status, opportunities, and future trends. *Big data and Cognitive Computing*, 5(1), p. 10.

19. Alex David, S., Varsha, V., Ravali, Y. and Naga Amrutha Saranya, N., 2022. Comparative analysis of diabetes prediction using machine learning. In *Soft Computing for Security Applications: Proceedings of ICSCS 2022* (pp. 155–163). Singapore: Springer Nature Singapore.

20. Ravikumar, S. and Kannan, E., 2022. Machine learning techniques for identifying fetal risk during pregnancy. *International Journal of Image and Graphics*, 22(5), p. 2250045.

21. Begum, A., Alex David, S., Hemalatha, D. and Kollipara, L. S. S., 2023. Deep learning-based lung cancer classification: Recent developments and future prospects. In *2023 International Conference on Advances in Computing, Communication and Applied Informatics (ACCAI)*, Chennai, India (pp. 1–8), doi:10.1109/ACCAI58221.2023.10200967.

22. Begum, A., Dhilip Kumar, V., Asghar, J., Hemalatha, D., & Arulkumaran, G. 2022. A combined deep CNN: LSTM with a random forest approach for breast cancer diagnosis. *Complexity*. https://doi.org/10.1155/2022/9299621.

23. Kannan, E., Ravikumar, S., Anitha, A., Kumar, S.A. and Vijayasarathy, M., 2021. Analyzing uncertainty in cardiotocogram data for the prediction of fetal risks based on machine learning techniques using rough set. Journal of Ambient Intelligence and Humanized Computing, pp. 1–13. https://doi.org/10.1007/s12652-020-02803-4.

24. Alex David, S., 2015. Patient tele-monitoring of vital parameters with emerging serve and virtual doctor patient using cloud server. *Global Journal of Pure and Applied Mathematics*, 11(4), pp. 1945–1951

25. Cong, L.W., Li, B. and Zhang, Q.T., 2021. Internet of things: Business economics and applications. *Review of Business*, 41(1), pp. 15–29.

26. Parihar, V., Malik, A., Bhawna, Bhushan, B. and Chaganti, R., 2023. From smart devices to smarter systems: The evolution of artificial intelligence of things (AIoT) with characteristics, architecture, use cases and challenges. In Bharat Bhushan, Arun Kumar Sangaiah, Tu N. Nguyen, *AI Models for Blockchain-Based Intelligent Networks in IoT Systems: Concepts, Methodologies, Tools, and Applications* (pp. 1–28). Cham: Springer International Publishing.

27. Matin, A., Islam, M.R., Wang, X., Huo, H. and Xu, G., 2023. AIoT for sustainable manufacturing: Overview, challenges, and opportunities. *Internet of Things*, 24, p. 100901.

28. Jiwani, N., Gupta, K. and Whig, P., 2023. Machine learning approaches for analysis in smart healthcare informatics. In Tawseef Ayoub Shaikh, Saqib Hakak, Tabasum Rasool, Mohammed Wasid, *Machine Learning and Artificial Intelligence in Healthcare Systems* (pp. 129–154). Boca Raton: CRC Press.

29. Johri, A., Bhadula, S., Sharma, S. and Shukla, A.S., 2022. Assessment of factors affecting implementation of IoT based smart skin monitoring systems. *Technology in Society*, 68, p.101908.

30. Ruotsalainen, P. and Blobel, B., 2023. Future pHealth Ecosystem-Holistic View on Privacy and Trust. *Journal of Personalized Medicine*, 13(7), p. 1048.

31. Malik, A., Parihar, V., Bhawna, Bhushan, B. and Karim, L., 2023. Empowering artificial intelligence of things (AIoT) toward smart healthcare systems. In Bharat Bhushan, Arun Kumar Sangaiah, Tu N. Nguyen, *AI Models for Blockchain-Based Intelligent Networks in IoT Systems: Concepts, Methodologies, Tools, and Applications* (pp. 121–140). Cham: Springer International Publishing.

32. Chang, Z., Liu, S., Xiong, X., Cai, Z. and Tu, G., 2021. A survey of recent advances in edge-computing-powered artificial intelligence of things. *IEEE Internet of Things Journal*, 8(18), pp. 13849–13875.

33. Divya, S., Panda, S., Hajra, S., Jeyaraj, R., Paul, A., Park, S.H., Kim, H.J. and Oh, T.H., 2022. Smart data processing for energy harvesting systems using artificial intelligence. *Nano Energy*, 106, p. 108084.

## Chapter 9

Sangeeta Sarangha

AI model for identifying Carnatic raga

*P. PadmaPriya Darshini, Mridula G. Narang,
H. Deepika, Aneesha M. Bail, and H. N. Sharul*

9.1 INTRODUCTION

Music is an activity that brings people together. It also holds a profound spiritual essence. The ancient Indians revered the power of music, giving rise to the realm of Indian classical music. Engaging with this art form demands unwavering dedication and a lifelong commitment from enthusiasts. However, the beauty of music lies in its versatility. Whether approached with reverence or casual enjoyment, it offers enriching experiences to individuals at any level of engagement.

Indian classical music stands as a tradition originating in South Asia, transcending geographical boundaries worldwide. Rooted in Vedic scriptures from over 6000 years, sacred chants evolved into a structured system of musical notes and rhythmic patterns. Embracing core elements like shruti (pitch) [1] swaras (tones for individual notes), raga (melodic framework), and tala (rhythmic cycle), this art form harmonizes closely with nature's rhythms. Drawing inspiration from phenomena such as seasons and times of day, it weaves musical modes known as 'ragas' and rhythmic cycles known as 'taals' enriched by improvisation, within the framework of established compositions and mathematical principles.

This creates a sense of spontaneity in the music, allowing each artist and performance to stand out as one of a kind.

9.1.1 Shruti

Shruti means 'to hear'. In Indian musical rhetoric, the term 'shruti' is in general used as 'microtone'. Shruti is actually defined as a particular frequency or pitch position in an octave. This is the main characteristic of a shruti [2]. Thus, Shruti alludes to the first pitch or oration selected by the performer. All the other following notes are given their position based on the very first frequency. The most crucial thing to note is that all the Indian musical notes are in regard to the relative starting position. This initial position of the notes in music is usually called the Shajama or Shadja that is

DOI: 10.1201/9781032618173-11

partly sung as a 'Sa'. It is only after establishing the position of this 'Sa' that we are now in a position to determine the rest of the other notes.

9.1.2 Swara

Swaras represent the music sea formed by combining powerful alphabets. Ragas are composed of consecutive groupings of swaras. 'Swaras' is the term used in Indian classical music for different notes that make up a scale or tune. In Western music, these notes could correspond to C, D, E, F, G, A, B. The swaras are essentially the seven basic notes or symbols of the pitch in classical music, namely, S(Sa), R(Re or Ri); G (Ga); M (Ma); P (Pa), D (Dha), N (Ni) [3,4]. Carnatic music has 16 swaras denoted as S, R1, R2, R3, G1, G2, G3, M1, M2, P, D1, D2, D3, N1, N2, and N3 [5].

Swara in Carnatic music is the primary kind of building block for melody harmony and rhythm. They serve as the ingredient of improvisation creation and execution borrowed from the Sanskrit word 'swara', meaning sound or note from the ancient Indian language. Each note stands for one point with respect to the range of pitches of an octave, and it is linked to a specific syllable. The Sa, which signifies the tonal or key, is located in one fixed place; other pitches adjoin it to make up a melody scheme, which is otherwise denominated as a raga. There are different types of swaras categorized through time factors, which stand for property notes and arohana–avarohana swaras, which means from ascending to descending, respectively. This system of classification helps us understand how a raga grows and what form it takes. Furthermore, in terms of rhythm, swaras are very important because every particular swara or musical note has its own time span or length of time for which it is uttered. The rhythmic aspect called 'tala' in the context of Carnatic music enhances the melodic constituents of the music to produce a musical continuum.

9.1.3 Raga

Indian classical music is based on the raga and its various characteristics. When we say melody we are referring to raga in Indian music. According to Ref. [5], raga consists of some specific notes that are characterized by special qualities like arohana, avarohana, and taal. As per the text, the classification of raga can be divided into two: linearity (all swaras) and nonlinearity (missing swaras) of arohana and avarohana.

Complete ragas with all seven notes are referred to as Melakarta ragas. Some examples of Sampoorna raga by variations are rishabha, gandhara, madhyama, panchama, daivatha, and nishadm. This constitutes only 72 ragas. The second group contains Janya raga. A number of Melakarta ragas and Janya ragas are included in Tables 9.1 and 9.2, respectively. The first row in Table 9.2 tells us the name of the Janya raga, followed by the swaras that occur in that raga.

Table 9.1 Sample Melakarta Raga

Ragam	S1	S2	S3	S4	S5	S6	S7	S8
Mayamalavagoula	S	R1	G3	M1	P	D1	N3	S
Kalyani	S	R2	G3	M2	P	D2	N3	S
Shankarabharana	S	R2	G3	M2	P	D2	N3	S
HariKhambhoji	S	R2	G3	M1	P	D2	N2	S
Kharaharapriya	S	R2	G2	M1	P	D2	N2	S

Table 9.2 Sample Janya Ragas

Janya Raga	Mohana	Hamsadhwani	Malahari
sa	sa	sa	sa
ri	ri2	ri2	ri1
ga	ga3	ga3	—
ma	—	—	ma1
pa	pa	pa	pa
da	da2	—	da1
ni	—	ni3	—
sa	sa	sa	sa
ni	—	ni3	—
da	da2	—	da1
pa	pa	pa	pa
ma	—	—	ma1
ga	ga3	ga3	ga3
ri	ri2	ri2	ri1

9.1.4 Applications

Here is an example of a system that shows artificial intelligence and machine learning could merge with conventional music territory to offer analysis tools, which would be useful for different reasons:

- **Music Education:** People studying or playing classical Indian music could get immediate comments about the usage of the right notes and scales, *Musicology Research*: These tools could be used by researchers to analyze vast amounts of information on music that would make it possible for them to study music history and music patterns development and evolution over time as well as application case studies of raga, *Archiving and Retrieval*: AI-powered tools can help convert and manage big repositories of Carnatic music, thus simplifying the search of songs based on particular musical properties.

9.2 LITERATURE SURVEY

The aim of Ref. [6] is to identify raga in Carnatic songs utilizing low-level signal traits. The process includes numerous steps such as signal separation, segmentation, extraction of characteristics, frequency mapping, employing a singer's voice, and subsequently raga identification. The system has modules on signal separation, segmentation, feature extraction, frequency mapping, and raga identification. The separation of the signal is vital in separating voice and music efficiently with an altered algorithm specifically for Carnatic music. Singer identification is employed when you want to find the fundamental frequency of the song, which is important for Carnatic music analysis. Frequency components are extracted from the signal and then directed to the swara sequence for determining the song's raga in the system. These frequency components and the fundamental frequency of the singer are used to identify raga, which in turn leads to its accurate identification. During the evaluation, three Melakarta ragas were successfully identified by the system for evaluation, using singers such as Nithyasree, M. S. Subbulakshmi, Balamuralikrishna, and Ilaiyaraaja. The system was able to exhibit segmentation points as well as frequency constituents of the recognized raga. In conclusion, below the low-level signal features and singer recognition, this chapter presents the use of a modern approach toward identifying ragas in Carnatic songs. A promising system has been formulated that can precisely identify ragas by analyzing the signal components' frequency and the fundamental frequency of the singer.

The aim of Ref. [7] is to use stochastic-based methods to analyze AlApana pieces of Carnatic music regardless of who sang them or the raga used. In Indian classical music, AlApana is a way of improvisation which helps to reveal the principle or feel (bhAva) of a raga by combining particular notes and phrases. This research identifies the shadja (base note), swaras (notes), and ragas in AlApana using stochastic models. The choice of the basic frequency by the singer for a melody is crucial in forming the music. The challenges of determining the AdhAra shadja because it changes with practitioners and practice are addressed in this chapter. The process entails examining the input sound file, estimating pitch frequencies, and transforming the pitch contour into a shadja determination density estimate. Semi-continuous Gaussian mixture models are employed to analyze the notes in the AlApana for identification of the swaras and ragas.

The main focus of Ref. [7] is on how likely each note interval is used by various raag composers, under the research scholars who tried to identify the swaras contained within AlApana and hence derived the correct raga simply from those notes. According to the stats, the present algorithm helps detect the starting swara for major, like 88.8%, AlApanas, which means that we can employ it as a tool for subsequent investigations; this concept is, however, not new, but it has been tried elsewhere, thus making its applicability limited here. While the research shows good accuracy in determining the tonic and notes, the accuracy for raga verification was quite modest at 62.13%. It notes the difficulties inherent in verifying ragas when some of their notes are absent or when they might show variations that would cause misclassification. The research on this is a good example of how statistics can be used to find basic aspects of southern Indian traditional music.

The objective of Ref. [8] is to develop a machine that can automatically recognize the raga in Indian classical music by investigating the sequence of notes sung and obtaining features that are connected to arohana–avarohana. The process needs the making of such strong persona in both individual swaras and their consequent sets, which would engender a set of 'different' swaras.

The program was tested after that on a huge dataset containing performances that are recorded raga-wise only. This resulted in the highest classification accuracy of 99.0% for the cross-validation tests and 75% for generalization experiments that were unseen. In addition, 20 distinct raga were tested using three to five tracks for each raga, leading to a 95% classification rate. A new methodology for extracting features from monophonic to polyphonic songs has been introduced. Arohana–avarohana is important for raga recognition in terms of the different musical structures in Indian Classical Music (ICM) that are more complicated than those found in other forms such as folk or Western music. The extraction tool went on to confirm that distance-based methods on chroma features have been able to capture more different songs than timbre clustering algorithms even though there is less variance so it can be seen why they are often applied separately.

The paper by Suma and Koolagudi [9] involves raga identification in Carnatic music without explicit knowledge of the scale by relying on low-level features like pitch and its derivative. The methodological feature of this research is the extraction of pitch contours from the songs through the auto-correlation technique as well as classification using a model of artificial neural networks (ANNs). The ANN model encompasses connected neurons that are responsible for processing presented characteristics needed to separate ragas. The system targets removing the demand both for explicit note recognition and for scaling relativity, concentrating on pitch changes defining the melody in each raga. The trial findings establish the efficiency of the recommended system planning for as early as 90% of correct identification based on mode IT as demonstrated in an overall analysis. The initial trial proved it all by recording 91.5% accuracy on average for traditional ragas while the five-fold validation technique led to 89.51% accuracy levels using music segments of different ragas. This demonstrates that, even when they have the same notes, there are two different basic ragas which, however, might consist of notes. This chapter specifies prospective evaluations, such as larger data sources filled with heterogeneous musical clips and research into areas like music transcription and raga-based recommendations.

Vignesh Ishwar and Ashwin Bellur explored the relevance of raga identification in Carnatic music as depicted by the paper [10] with machine learning techniques that can be used to analyze ragas in Carnatic music.

Motifs are defined as prosodic phrases that have specific phonological properties, with respect to their pitch height, pitch shape, rhythmic aspects, and envelope commence. In recognizing ragas, motifs are vital because they retain exclusive melodic designs and attributes for each raga. Managing to recognize a popular raga involves compiling specific concert collections; this is done to achieve phrases which depict the kind of melody associated with that particular raga which is there.

Hidden Markov models (HMMs) are selected because they distinguished ragas according to motifs, giving due consideration to changes and subtleties found in Carnatic music. The authors applied continuous density HMMs for modeling motifs, where HMM structures were designed to match the particular features of each motif. It was proved by the research findings that there is only one raga per each motif—showing how worthwhile it is to study ragas using motives. The study obtained information on the intricate melodic structures of ragas in Carnatic music through the use of machine learning techniques like HMMs.

In summary, Ishwar and Bellur examined why motivic analysis and machine learning are important for recognizing ragas in Carnatic music. The study demonstrated a new way to recognize raga in Carnatic music by analyzing motifs and using HMMs, taking into consideration that Carnatic music traits are subtle and vary, thereby broadening its scope.

9.3 OVERVIEW OF SANGEETA SARANGHA

The overarching goal of Sangeeta Sarangha is to develop a comprehensive system for analyzing and identifying raga in Carnatic music. Figure 9.1 explains the workflow of processing a Carnatic song to recognize basic musical components such as swaras and raga. The process can be summarized as follows.

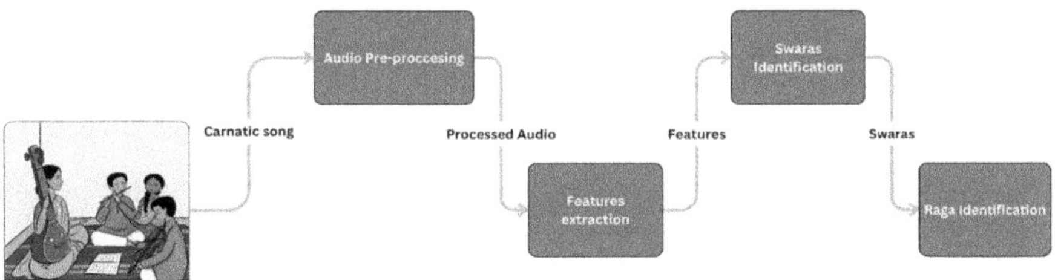

Figure 9.1 Overall block diagram of Sangeeta Sarangha.

Carnatic music is one of the ancient Indian classical music forms known for its use of complex ragas and talas. The input to Sangeeta Sarangha is a Carnatic song. Before analyzing the song, it must be pre-processed, which involves normalizing volume levels, eliminating noise, and possibly converting it into a suitable format for further processing. The main goal is to improve the audio signal quality/strength so that characteristics can be extracted more accurately and with fewer errors. The pre-processed audio signal is then prepared for detailed analysis. It serves as the intermediary between the raw audio input and the retrieval of meaningful musical features. Custom algorithms are applied to extract sound elements such as pitch and rhythm. The next step is swara identification. In Indian classical music, swaras are similar to the Western concept of solfege and are considered musical notes. Ultimately, using the identified notes, the system attempts to ascertain the raga of the song.

9.3.1 Modules of Sangeeta Sarangha

Sangeeta Sarangha has modules for swara and raga identification, raga database creation, and web application development. Detailed explanations of all these modules are as follows:

- **Algorithm for Swara Identification:** Development of an algorithm that can detect Swara in Indian classical music involves creating and executing a basic musical note identification algorithm in the primary place. Fundamental frequency analysis is utilized by this algorithm to ascertain the individual swara notes from input audio files. Other methods used include pattern recognition, which involves the analysis of frequency and period information for each swara as found within the sample sound.
- **Creation of Raga Database:** After the swaras are recognized, the subsequent task is to gather a database of the traditional raga patterns. Every raga has its own specific way of organizing swaras, which is referred to as its characteristic melodic structure. This database holds data on the swara sequences that are the basis of each raga and the ratio of each swara to the fundamental Sa. This has been shown in Figure 9.2.
- **Algorithm Development for Raga Identification:** With the swara identification algorithm and raga database, the next phase is to create an algorithm that can recognize particular raga patterns. This algorithm

```
{"A#": 116.5409, "A": 110.0, "B": 123.4708, "C#": 69.2957,
"C": 65.4064, "D#": 77.7817, "D": 73.4162, "E": 82.4069,
"F#": 92.4986, "F": 87.3071, "G#": 103.8262, "G": 97.9989}
```

Figure 9.2 Fundamental frequencies of the shruti to find the pitch of the singer.

will examine the distinct arrangement of swaras in a specified audio file and will match it with similar patterns in the database. Through the match of the swara sequence extracted to the entries of the database, the algorithm can find the most probable raga that is connected to the input audio.

- **Web Application Development**: Raga identification will be a simple and user-friendly process since the website development will be done for it. This application is going to be one based on a graphical user interface, which is composed of simple navigation and interaction. The users will be allowed to upload the audio files themselves to the application, and the application will utilize the developed algorithms to analyze the music and to find the raga.

9.4 RESULTS AND DISCUSSION

This subsection discusses the evaluation process of Sangeeta Sarangha.

Step 1: *Collecting Swara Frequencies*

Firstly, swaras (Sa, Ri1, Ri2, Ga3, Ma1, Pa, Da1, Da2, Ni3) are recorded and saved. Their frequencies are calculated with the help of the Librosa library. The result is a list of frequencies for each swara.

Step 2: *Clustering Frequencies*

A list is created to cluster the frequencies. To make this list's grouping rational and systematic, frequencies were allocated to these sub-lists based on their closeness to the sub-group average. More specifically, any difference between the mean central value of a sub-group and some of it is less than one; a frequency would be conversely augmented. To some extent noise reduction is enabled through this method. The process gives rise to a list populated by various sub-lists functioning as groups.

Step 3: *Weighted Mean Calculation*

In this calculation, the weighted mean for each cluster in the list is determined. The weight is assigned based on the length of the cluster, and the value is the mean frequency. Consequently, the recorded value represented the frequency of the swara.

Step 4: *Normalization and Standardization*

Once the frequency of all swaras was determined relative to the frequency of Sa, a list of normalized values was created. From the average of these normalized values, each respective swara would have one common ratio. The ratio is the same for all the users.

Step 5: *Ratio Storage*

A dictionary was used to store the ratios which had the swaras as their keys and the final ratios of each swara as their respective values. The swara ratio is presented in Figure 9.3.

{"sa": 1.0, "ri1": 1.0502930852800931, "ri2": 1.125557443257559,
"ga3": 1.253607019907006, "ma1": 1.3325629015795974, "pa": 1.4954337412417669,
"da1": 1.5774395457015526, "da2": 1.687713914863785, "ni3": 1.8822683388048216}

Figure 9.3 Swara ratios.

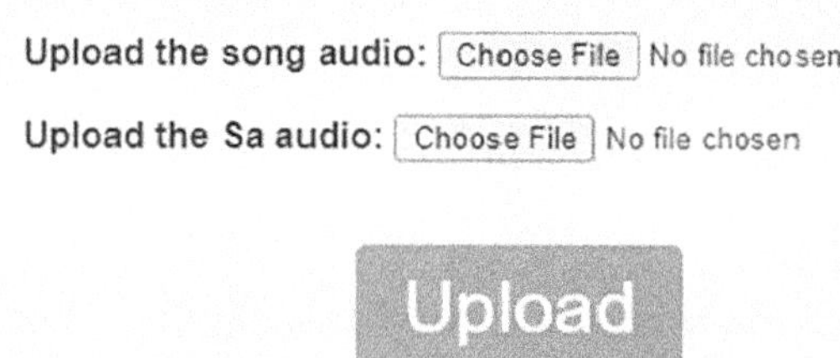

Figure 9.4 Graphical user interface to upload a song and 'SA' swara of the user.

Step 6*: Swara Extraction*

Users upload their song along with the singer's Sa frequency, which will help determine the pitch at which the song is singing. The audio processing is done using the Librosa library that extracts swaras. By using the value of Sa, we get all the expected swara frequencies of the singer. Figure 9.4 shows the user interface for uploading files.

Step 7*: Matching Frequencies*

Iteration through the frequencies of the whole song to discover the swaras closely matches with one of the expected swaras. The count of the swaras is stored in a dictionary.

Step 8*: Raga Identification*

A dictionary was used, whose keys were ragas, and values were associated lists of swaras appearing in that raga, as shown in Figure 9.5. The raga of the song would be identified by dividing the count of swaras belonging to a raga that appeared in the frequency matching, by the count of all the swaras observed in the song.

The result would be a matching percentage that shows the similarity of the song to a particular raga. The song is in close resemblance to the raga if its percentage is high, and a sample is shown in Figure 9.6.

```
{
    "Mayamalavagoula":["sa","ri1","ga3","ma1","pa","da1","ni3","sa1"],
    "Mohana":["sa","ri2","ga3","pa","da2","sa1"],
    "Hamsadhwani":["sa","ri2","ga3","pa","ni3","sa1"],
    "Shankarabharana":["sa","ri2","ga3","ma1","pa","da1","ni3","sa1"]
}
```

Figure 9.5 Sequence of swara corresponding to each raga.

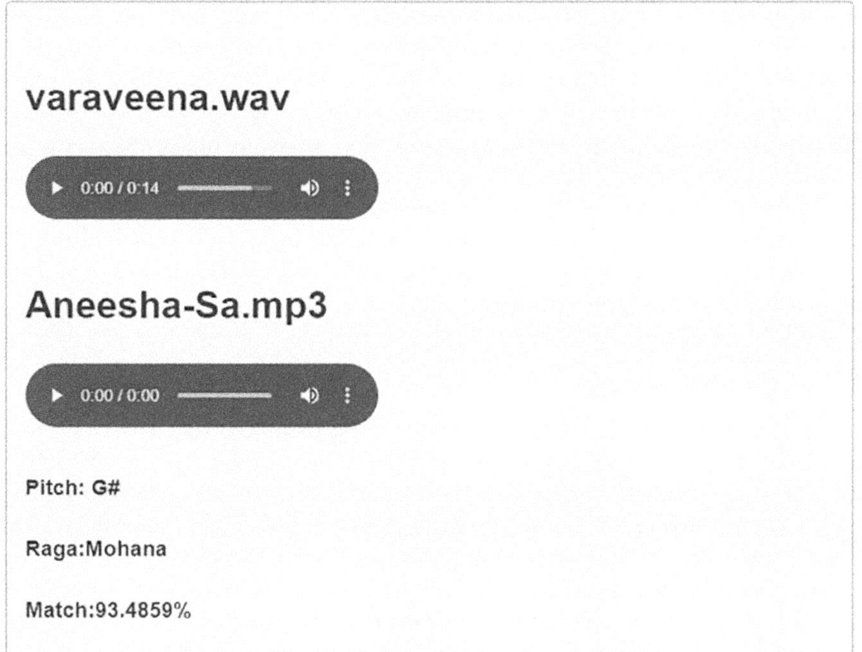

Figure 9.6 Identified raga of the given song.

9.5 CONCLUSION AND FUTURE SCOPE

Sangeeta Sarangha can detect swaras and ragas from audio recordings using digital signal processing techniques. Utilizing the Librosa library for frequency extraction and incorporating clustering techniques to purify these frequencies, we have developed a method that reduces noise and standardizes swara frequencies. Accurate identification of swaras is ensured across different singers regardless of their singing pitch through the use of standardized ratios. Furthermore, the system can map these swaras to known ragas, providing a similarity measure that indicates the closeness of songs to a particular raga in percentage terms. This implies improved digital systems for analyzing music, which would be helpful to musicians and musicologists working on Indian classical music issues.

9.5.1 Limitations

Even though it is successful, Sangeeta Sarangha still has a number of constraints. With the quality of initial audio recordings being critical for accuracy during swara extraction and classification, performance may be imperiled by recording equipment quality as well as background noise. Additionally, the interpretation of musical notes through grouping together in a clustered manner is quite simplistic even though this method helps eliminate the noises in terms of on-stage vocal pitch changes. Also, the existing model depends mainly on a small number of ragas, which could mean that it may not do as well with rarer or more difficult ragas. Additionally, because it uses a monolithic architecture, it does not allow for independent scaling of swara matching algorithms or raga identification modules and may not scale well or be easily maintained. In monolithic architecture, it's difficult to scale out on specific components such as the swara matching algorithm or the raga identification module.

9.5.2 Future scope

In a production environment, it would be best to switch to a micro-services architecture as this would greatly benefit the project. By design, in such an architecture it would be possible to scale different services autonomously, for instance, storage of data, swara extraction, extraction of swaras, and raga identification. When the number of users uploading songs is more than usual, just the swara extraction and user interaction services should be scaled but not other parts of the system. It is intended to help make things go even more quickly both for developing applications and for the installation of new releases using mechanisms like micro-services where separate groups can work on components separately. The change of one function would not have to mean that other parts have waited in it. Moreover, it boosts system reliability and availability in general by preventing the collapse of an entire system in case a single micro-service fails. Another benefit of micro-services is their technology diversity because each of them can be created with the most appropriate technology to meet the specific demands and needs that any particular service requires, hence making it possible for developers or businesses to select the appropriate tools they need for specific tasks.

Also, the development of capabilities for real-time swara and raga analysis during the real-time performances can indicate to the musicians what they need to do next and improve interactive learning tools. Integrating more advanced machine learning models could improve accuracy in swara and raga identification, especially in handling variations observed during live performances. Moreover, developing capabilities for real-time swara and raga analysis during live performances could provide instant feedback to musicians and enhance interactive learning tools, while integrating more advanced machine learning models could improve accuracy in swara and raga identification, especially in handling variations in live performances.

REFERENCES

1. Sunadam (2016) *Introduction to Shruti: The Foundational Pitch.* Available at: https://www.sunadam.com/2016/03/14/introduction-to-shruti-the-foundational-pitch/
2. Vidwans, V. (2016) *The Doctrine of Shrutis in Indian Music.* Pune: Flame University. Available at: http://www.indian-heritage.org/music/The Doctrine ofShrutiInIndianMusic-Dr.VinodVidwans.pdf
3. Raag Hindustani (n.d.) *Introduction to Hindustani Classical Music.* Available at: https://raag-hindustani.com/Introduction.html
4. MakingMusic(n.d.)*IntroductiontoIndianClassicalMusic.*Availableat:https://www.makingmusic.org.uk/resource/introduction-indian-classical-music
5. Katte, T. (2013) 'Multiple techniques for raga identification in Indian classical music', *International Journal of Electronics and Computer Engineering*, 4(6), pp. 82–87.
6. Sridhar, R. and Geetha, T.V. (2009) 'Raga identification of Carnatic music for music information retrieval', *International Journal of Recent Trends in Engineering*, 1(1), p. 571.
7. Ranjani, H.G., Arthi, S. and Sreenivas, T.V. (2011) 'Carnatic music analysis: Shadja, swara identification and raga verification in alapana using stochastic models', in *2011 IEEE Workshop on Applications of Signal Processing to Audio and Acoustics (WASPAA)*. IEEE, pp. 29–32.
8. Shetty, S. and Achary, K.K. (2009) 'Raga mining of Indian music by extracting arohana-avarohana pattern', *International Journal of Recent Trends in Engineering*, 1(1), p. 362.
9. Suma, S.M. and Koolagudi, S.G., 2015. Raga classification for carnatic music. In Information Systems Design and Intelligent Applications: Proceedings of Second International Conference INDIA 2015, Volume 1 (pp. 865-875). Springer India.
10. Bellur, A., Ishwar, V. and Murthy, H.A., 2012. Motivic analysis and its relevance to raga identification in carnatic music. In Serra X, Rao P, Murthy H, Bozkurt B, editors. Proceedings of the 2nd CompMusic Workshop; 2012 Jul 12-13; Istanbul, Turkey. Barcelona: Universitat Pompeu Fabra; 2012. p. 153-157. Universitat Pompeu Fabra

AIoT-driven advances in personalized healthcare monitoring and diagnostic systems

Rajalakshmi N. Nagarnaidu Rajaperumal

10.1 INTRODUCTION

The COVID pandemic intensified the strain on global healthcare systems, further burdening facilities already managing chronic illnesses and an aging population. Healthcare 5.0 offers a new era of Internet of Things (IoT) systems powered by AI. Artificial intelligence (AI) systems have quite ingeniously decreased the need for repetitive human labor and have the potential to produce outcomes. They offer intelligent self-management, personalized medicine, intelligent remote monitoring, and emotive telemedicine as a part of ubiquitous care by utilizing cutting-edge technologies within an AIoT framework regardless of people's location or other limitations. AI technology is an incredibly fascinating purpose of life and work. It is just about "choosing" the correct decision at the right moment. AI is the science of infused machines with intelligence to do jobs that have historically needed human intelligence. AI-based systems are rapidly evolving with respect to usefulness, versatility, efficiency of processing, and capacities. Less routine jobs are becoming more and more capable of being performed by machines.

The combination of AI with IoT has been a major focus of research in the healthcare industry, as illustrated in Figure 10.1. The IoT is a network of physical objects, equipment, and sensors that communicate and share data. IoT is a broad term that includes an excessive number of sensors, actuators, data processing, storage, and networking capabilities linked via the Internet. The IoT allows devices to detect their environment and transmit, store, and analyze data, all of which rely on the processing stage. Wearable technology and small sensors on phones and tablets are driving AIoT development due to their compact size, ease of use, and deployment. These devices track user health metrics, monitor patient surroundings, and ensure treatment plan compliance, making them the primary technology driving AIoT development. Actuating devices assist users carry out an action. Examples of these include robots that assist individuals in getting out of bed and smartphone apps that alert diabetics when their blood sugar is low. Actuators are typically found in systems that have several

DOI: 10.1201/9781032618173-12

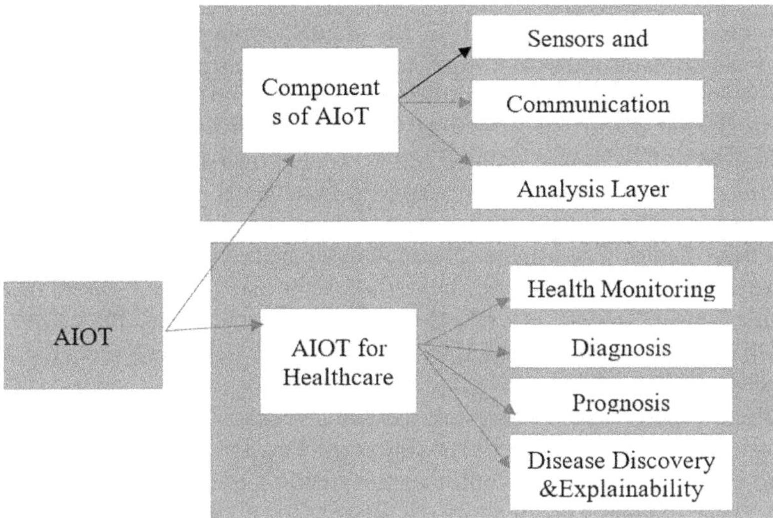

Figure 10.1 AIoT components and application.

perceptual devices. IoT technologies are increasingly being utilized in the healthcare industry to improve patient outcomes and care, offering new opportunities for personalized treatment plans, remote monitoring, and efficient healthcare delivery. Fortino et al. (2014) and Rajalakshmi et al. (2023) said that IoT technologies simplify healthcare operations by automating administrative duties. This lowers the need for a large administrative staff and saves time, all of which contribute to cost savings. Additionally, routine operations like medication dispensing and vital sign monitoring can be automated to reduce error risk and labor costs.

The primary way IoT reduces healthcare costs is by preventing costly medical complications, shortening hospital stays, and minimizing the need for expensive interventions through early detection and ongoing monitoring. Furthermore, it can promote patient participation by giving consumers greater control over their care, allowing them to better monitor their health information and connect with medical experts. Vital signs are tracked by wearables and apps, and medical records can be accessed through patient portals. Telemedicine and remote monitoring enable real-time data sharing and virtual appointments, while mobile apps streamline data management and sharing. Secure messaging platforms facilitate direct interactions with healthcare professionals, and IoT devices automatically transmit data. To guarantee that users can operate these technologies with confidence, education and training are crucial. Patients who utilize these services actively participate in their healthcare, which fosters improved communication and engagement and, ultimately, improves health outcomes.

The electrocardiogram (ECG) and photoplethysmogram (PPG) are wearable sensors that are used for the identification of cardiac abnormalities, prediction of cardiac events, and cardio-respiratory health monitoring. Utilizing one or more electrodes, the ECG captures and analyzes the heart's electrical activity, providing valuable insights into its overall health. Meanwhile, PPG sensors employ light directed into an artery to gauge significant signs like blood oxygen saturation and heart rate. As blood absorbs some of the light, the rest either passes through or reflects, enabling comprehensive health monitoring. Heart activity waveforms and vital signs are recorded by measuring the non-absorbed light using a photodiode or comparable sensor. It detects arrhythmias and heart disease. PPG and ECG sensors are also easily accessible as built-in or add-on components for Apple watches and fitness devices.

Pulse oximeter sensors measure oxygen levels and blood pressure during surgeries, Chronic Obstructive Pulmonary Disease (COPD), and asthma. Blood pressure monitors monitor hypertension and hypotension. Research is underway on using electroacoustic sensors to collect sound-based cardiac activity patterns, improving wearability. These sensors use cutting-edge materials and production processes to construct strain and high-sensitivity pressure sensors. Electromechanical sensors, exemplified by strain and highly sensitive pressure sensors, rely on materials that exhibit changes in capacitance or electrical resistance when subjected to strain or pressure. These sensors serve to measure the heart rate and blood pressure, two critical indicators, which have been the subject of numerous studies. A glucose sensor detects diabetes by monitoring blood glucose and insulin levels. A temperature sensor diagnoses hypothermia or fevers by measuring body temperature. A respiratory sensor monitors breathing abnormalities such as sleep apnea by measuring respiratory rate and rhythm. Several investigations employed tools such as inertial measurement units (IMUs) and strain sensors to assess changes in pulmonary pressure during breathing. IMUs use accelerometers and gyroscopes to collect respiratory data, while strain sensors track chest wall movement by detecting material impedance variations during inhalation and exhalation cycles. The use of mechano-acoustic sensors to extract respiratory information from respiration sounds has also been increasingly investigated. The measurement of thoracic impedance changes during the respiratory process using electrodes attached to the chest.

AI is being integrated into IoT devices in healthcare to enhance predictive analytics for patient monitoring. Most devices have minimal processing power, allowing only light AI algorithms to be implemented. These devices collect huge amounts of patient data; the collected raw data may be pre-processed and analyzed using machine learning (ML) techniques. AI can generate predictive models to forecast potential health issues before they occur, triggering alerts and notifications for healthcare providers or patients. AI can also provide personalized insights and recommendations based on individual patient data, improving overall health outcomes and disease management. As more data is collected and analyzed, AI algorithms can continuously learn and improve their predictive capabilities. This integration also enables remote monitoring and telemedicine, facilitating proactive healthcare management for patients with chronic conditions or those living in remote areas. The synergy of AI and IoT technology enhances healthcare monitoring and diagnostics. AIoT components and applications are shown in Figure 10.1.

10.2 PERSONALIZED MEDICINE

Personalized medicine utilizes a range of sensors to collect accurate, individualized data on a person's health behaviors and environment, enabling tailored healthcare monitoring as shown in Figure 10.2. Sensors in the physical layer of IoT architecture are playing a crucial role in personalized medicine by collecting comprehensive health data for individual patients. These sensors capture physiological parameters, behaviors, and environmental factors, providing a holistic view of a patient's health status. Healthcare providers utilize vital signs monitoring, activity tracking, blood glucose and oxygen levels, and environmental sensors to understand patients' health status and develop personalized treatment plans. The integration of sensor data enables healthcare professionals to provide a comprehensive view of patients' health, enabling precise diagnosis, proactive disease management, and personalized treatment interventions. Continuous monitoring is crucial for detecting health issues early and implementing timely interventions, leading to improved health outcomes and better patient care.

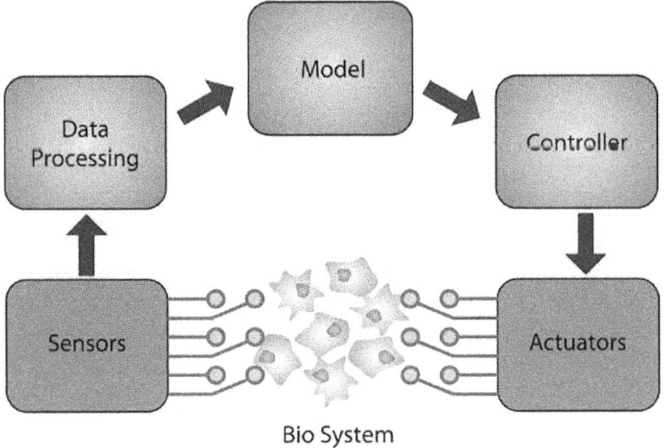

Figure 10.2 Bio-system.

A bio-system is a healthcare framework that uses biological knowledge, advanced technologies, and patient-centric care to revolutionize healthcare delivery. It considers each patient's distinct genetic profile, lifestyle choices, and environmental factors. The system collects comprehensive health data from various sources, using advanced data processing and AI models to extract actionable insights. Personalized treatment plans are tailored to the patient's health profile and genetic makeup, guiding medication selection and dosage adjustments. Patient engagement and shared decision-making are fostered through personalized health information, educational resources, and decision-support tools. Actuators and controllers are essential components in personalized medicine, enabling precise and targeted interventions tailored to individual patient needs. By combining advanced technologies, biological insights, and personalized care models, bio-systems enable healthcare providers to deliver optimized treatments, enhance patient outcomes, and improve the quality of life in personalized medicine.

10.2.1 Data collection

Patients generate a vast amount of raw data through IoT devices and sensors. This includes vital signs, medication adherence, activity levels, and sometimes genetic information. Biometric sensor measures physiological parameters such as blood pressure, heart rate, body temperature, and respiratory rate. Devices like fitness trackers, smartwatches, and wearable patches often incorporate biometric sensors. Biosensors detect specific biological analytes like glucose, hormones, proteins, or DNA. They are analytical tools that identify disease biomarkers by using biological recognition components like antibodies and receptors. Transducers are then used to quantify the biomarkers. Any material, structure, or mechanism that can be detected in the body or its by-products that affect or forecast the occurrence of an event or illness is a biomarker. This comprises proteins, some metabolites, hormones, and nucleic acids.

In the field of medicine, biomarkers are becoming more and more significant, especially in personalized medicine. They are useful in dose selection and prognostic prediction. Additionally, they might be useful for patient stratification based on safety or efficacy predictions, as well as for identifying therapeutic and unfavorable responses. Thus, biomarkers play a crucial role in enabling personalized medicine or "delivering the appropriate medication to the appropriate patient at the appropriate time in a dosage," by helping to identify which patients are suitable for treatment with particular medications. Researchers are using advanced equipment to analyze high-dimensional biomarker panels to identify disease endotypes in various clinical presentations.

This is really helpful for precision medicine in treating complicated human illnesses. Biosensors are used in devices for monitoring blood sugar levels in diabetes, detecting biomarkers in cancer, or assessing various health conditions. To identify prognostic biomarkers that can be related to death, worsening of the disease, and more severe illness, researchers usually conduct studies that are specific to the disease, such as regular imaging collection of data (Ciurtin et al., 2019), blood, urine, tears, and even breath (Robinson et al., 2020; Coelewij et al., 2021; Glazyrin et al., 2020; Torok et al., 2013; Sola Martínez et al., 2020). Omics analysis of such biological material is used to study disease activity prediction and diagnosis in inflammatory chronic diseases. Examples of this type of analysis include metabolomics, proteomics, RNA sequencing (also referred to as "big data"), and autoantibody data. Genomic sensor analyzes an individual's genetic makeup, helping to understand genetic predispositions to diseases and guiding personalized treatments based on genetic profiles. Environmental sensors track environmental factors such as UV exposure, temperature, air quality, humidity, and pollution. They can be integrated into wearable devices or home monitoring systems to assess their impact on an individual's health. Technologies like MRI, CT scans, ultrasound, and PET scans capture detailed images of internal body structures, aiding in disease diagnosis, monitoring treatment efficacy, and predicting health outcomes. Activity and motion sensors detect movement and activity levels, providing insights into physical activity patterns, sleep quality, and mobility. They're often integrated into wearable devices. Chemical sensors measure chemical compositions in bodily fluids or environmental samples, aiding in diagnostics, drug monitoring, or detecting specific substances. Remote monitoring sensors enable remote monitoring of patients, facilitating continuous healthcare without requiring physical presence. They can include telemedicine devices, remote patient monitoring systems, or IoT-based health monitoring tools. The integration of sensors enables healthcare professionals to collect comprehensive data, enabling personalized treatments and interventions based on individual needs and characteristics.

10.2.2 Data analysis

AI algorithms can significantly enhance the personalization of treatment plans by analyzing patients' responses to medications and tracking their health progress. These AI systems sift through data to uncover patterns and relationships, such as how a patient's physiological responses correlate with specific medications or treatment approaches. ML and AI are key technologies, with ensemble learning enhancing model generalization and accuracy by combining multiple computational models into a single prediction. Deep learning is particularly helpful when working with unstructured data, such as speech, text, and images. It does this by using artificial neural networks, a formalization based on the human brain, to identify patterns or associations in the data. Based on medical records containing over 800 predictors, such as blood tests, wound parameters, and demographic traits, ML is utilized for the early diagnosis of chronic wounds. ML models predict diagnostic diseases like Alzheimer's disease, cardiovascular diseases, and atopic dermatitis using 86 laboratory tests, using a multi-class model. AI is a powerful tool that combines structured and unstructured data to make semantic-level decisions, aiding health professionals in making better decisions. This idea of shared expertise highlights the collaboration between AI systems and human experts, resulting in changes to the workforce and the need for new skills. Nonetheless, advanced AI models and top-tier business applications demand skilled professionals who have access to the latest technology.

10.2.3 Genomic data analysis

As a new area of precision medicine research, radio genomics aims to predict a patient's likelihood of experiencing harm from radiation therapy by identifying correlations between gene expression and medical imaging characteristics. The radio genomic analysis of any disease can be performed using deep learning techniques. Hamidinekoo et al. (2018) proposed breast cancer classification using a deep learning technique. They suggested a computer-based method for modeling breast cancer and creating a feature/phenotype mapping between mammographic abnormalities, and their histological representation is the Mammography–Histology–Phenotype–Linking Model. Deep learning architectures have revolutionized medical image analysis, achieving unprecedented performance in tasks like tissue segmentation, classification, and clinical outcome prediction. In particular, three distinct methods for deep learning: off-the-shelf deep features, transfer learning, and training from scratch. When sufficient training in medical image data is available, the simplest method for developing deep neural networks is to train them from scratch. The study by Huang et al. (2019) introduced survival analysis learning with multi-omics neural networks (SALMON), a method that combines gene expression data and cancer biomarkers for prognosis prediction. SALMON used six combinations of multi-omics data as input sources, including 57 features of mRNA-seq data; 12 features of miRNA-seq data; 69 features of mRNA and miRNA integration; 71 features of mRNA, miRNA, copy number burden, and tumor mutation burden; 72 features of mRNA, miRNA, demographical, and clinical data; and 74 features of tumor mutation burden. SALMON was tested against multiple alternative survival prognosis algorithms, including Cox-nnet, DeepSurv, GLMNET, and RSF. The study identified multiple biological processes and validated specific mRNA-seq co-expression modules critical to breast cancer prognosis. Through additional patient age stratification, SALMON verified that distinct age groups possess distinct primary characteristics that influence survival prognosis outcomes.

However, the number of training medical images is not enough for training from scratch, which leads to overfitting. The transfer learning approach uses a pre-trained neural network as a feature extractor on natural image data and then trains it using specific medical image data. The extracted features are then used to train a traditional classifier (Support Vector Machine) to classify the glioma characterization, head and neck cancer classification, and fatty liver grading. Additionally, a framework of multiple residual convolutional neural networks (CNNs) is being suggested for noninvasively predicting isocitrate dehydrogenase genotypes in gliomas.

Moeskopset al. (2016) proposed brain MRI tumor segmentation, Joyce et al. (2018) suggested whole heart segmentation, Hoogi et al. (2017) proposed liver tissue segmentation, and Cai et al. (2016) gave MRI pancreas segmentation. Because MRI lesions vary in size, pre-trained networks like GoogLeNet, DenseNet, and VGGNet must be able to operate at multiple scales. The Inception module, which consists of three convolution layers with various kernel sizes, is the fundamental component of GoogLeNet. The features of patterns with varying sizes can be captured using these kernels with a far lesser learning rate (0.0001).

Utilizing transfer learning involves initially pre-training a DenseNet201 model on a dataset of natural images, followed by refining its performance using a dataset of medical images. The DenseNet architecture is given in Figure 10.3. The adaptability of the DenseNet201 model allows for fine-tuning pre-trained network weights on smaller datasets and training network weights on more extensive datasets. Dense blocks within the model are specifically designed for down-sampling and are structured with layers referred to as transition layers, comprising batch normalization, a 1×1 convolution layer, and a 2×2 average pooling layer. The growth rate hyperparameter, denoted as m in DenseNet201, serves as a key determinant of the model's success, showcasing the efficacy of its dense architecture. The unique design, treating feature maps as the network's global state, allows DenseNet201 to excel even when employing lower growth rates. This distinctive structure ensures that all feature maps from preceding levels remain

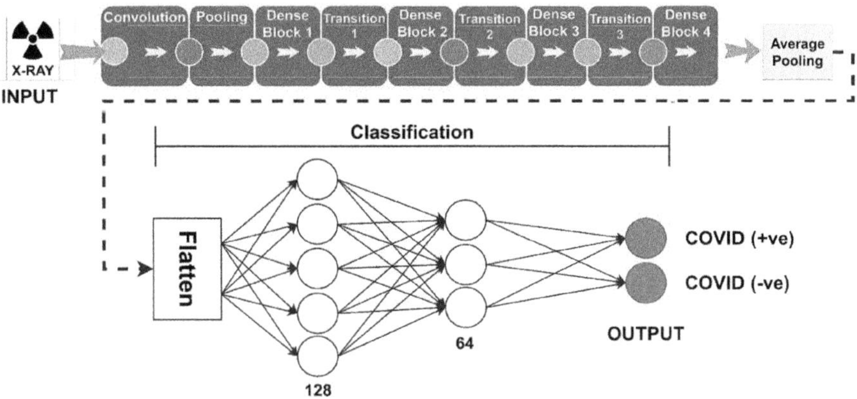

Figure 10.3 DenseNet architecture.

accessible to each subsequent layer. With each layer contributing m feature maps to the global state, the cumulative number of input feature maps at the nth layer is precisely defined as follows:

$$\text{Feature map} = m + m(n-1) \tag{10.1}$$

Here, m signifies the channels in the input layer. Illustrated in Figure 10.4, a 1×1 convolution layer precedes each 3×3 convolution layer, enhancing computational efficiency by diminishing the input feature map count, which typically exceeds the output feature maps denoted as m. The classification process, as depicted in Figure 10.4, involves two flattened layers equipped with 128 and 64 neurons, respectively.

Trivizakis et al. (2019) introduced an innovative three-dimensional (3D) CNN specifically designed for distinguishing primary liver tumors from metastatic liver tumors using diffusion-weighted MRI (DW-MRI) data. This 3D CNN was developed for seamless tissue classification in 3D tomographic data without necessitating additional preprocessing steps such as cropping, annotation, or region-of-interest detection. The proposed network comprises four consecutive 3D convolutional layers with a stride of 3, utilizing a kernel size of $3\times3\times3$ and rectified linear units as the activation function. Batch normalization is applied after the second and fourth layers. A fully connected layer with 2048 neurons, featuring a 50% dropout, and a SoftMax layer were incorporated for binary cancer classification using a dataset consisting of 130 DW-MRI scans, meticulously employed for training and validation. The DW-MRI data at b1000 served as the training samples. To enhance the model's robustness, data augmentation techniques were employed, including elastic deformation with a 9% alpha affine, 90° and 270° rotations, flipping both vertically and horizontally, adjustments in brightness, and mirroring with eight-pixel wide reflections. The 3D CNN network underwent training with input data of size $30\times256\times256$ over a span of 20 epochs. As showcased in Figure 10.5, the model demonstrated an impressive testing accuracy of 85.5% using SoftMax for classification. Figure 10.5 provides a visual representation of the 3D CNN, highlighting its distinctive features, particularly the four consecutive strides.

10.3 PERSONALIZED INSIGHTS

AI-driven healthcare has significantly improved treatment plans and patient outcomes. It has been used in

- Cancer treatment,
- Cardiovascular disease management,
- Diabetes care,
- Mental health support, and chronic disease management.

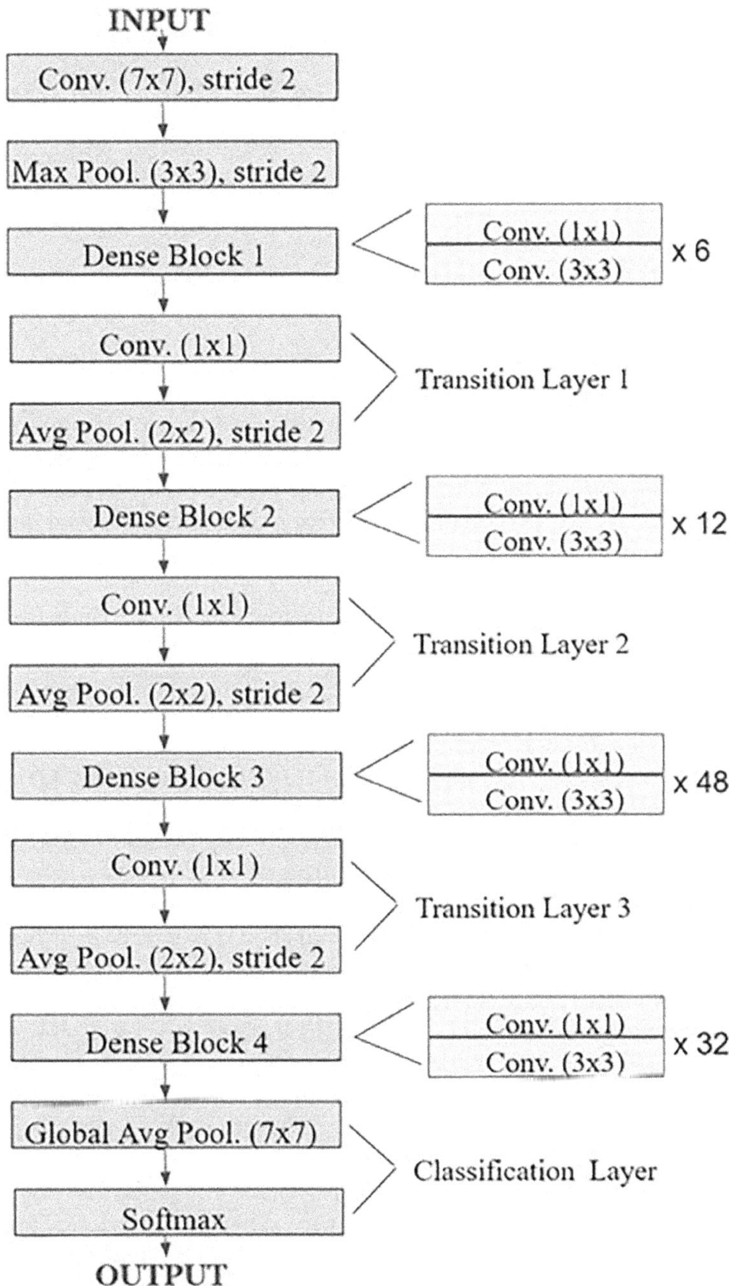

Figure 10.4 DenseNet201 layer.

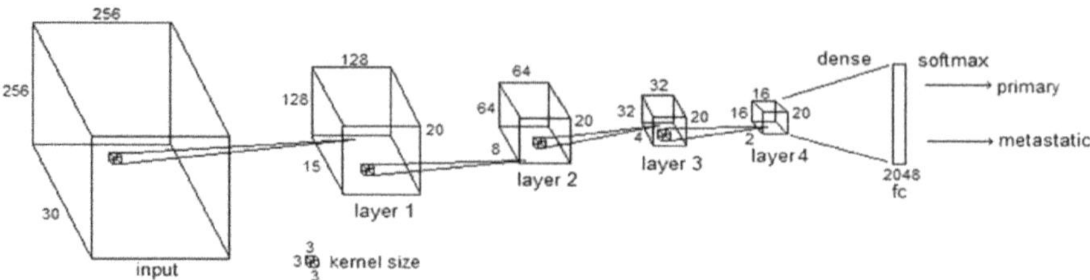

Figure 10.5 3D CNN featuring four-stride 3D convolutional layers.

AI algorithms analyze genetic information to design targeted therapies, predict heart disease, create personalized insulin delivery plans, and provide mental health support. AI-powered apps also help manage chronic diseases by monitoring patient data and suggesting personalized treatment plans. These advancements demonstrate the transformative potential of AI in delivering personalized healthcare solutions, improving patient outcomes, and enhancing healthcare system efficiency.

Based on the analysis, AI can generate personalized insights about a patient's response to medications. It can identify which treatments or dosages are more effective for specific individuals, considering factors like genetics, lifestyle, and other health conditions. Using these insights, healthcare providers can adapt and customize treatment plans for each patient. AI helps in recommending adjustments to medications, dosages, or therapies to optimize effectiveness and minimize side effects. Tailored treatment recommendations can be derived through the exploration of interactions between a patient's covariates and treatment effectiveness, leveraging a deep neural network based on Cox proportional hazards. Katzman et al. (2018) introduced the Cox proportional hazards DeepSurv Network, a deep feed-forward neural network designed to predict how a patient's covariates influence their hazard rate, dictated by the network's weights (θ). DeepSurv comprises essential elements, featuring a dropout layer and a fully connected layer. The patient's baseline data x serves as the input for the network, navigating through fully connected layers and dropout layers constituting the network's hidden layers. The network's output is a single node with linear activation, approximating the log-risk function in the Cox model, and training focuses on optimizing the average negative log partial likelihood with regularization, serving as the objective function. This methodology ensures that the deep neural network captures the intricate relationships between covariates and treatment effectiveness, paving the way for personalized treatment recommendations.

10.4 CONCLUSION

AIoT, the integration of AI and IoT, has revolutionized personalized medicine by improving data collection, enabling tailored treatment plans, remote patient monitoring, and early health detection. This approach focuses on the unique genetic makeup and biological characteristics of each patient, minimizing adverse effects and empowering them. This patient-centric model fosters a collaborative relationship between healthcare providers and patients, transforming healthcare delivery into an informed, participatory, and effective partnership. As we continue to unlock the mysteries of the human genome, the potential for personalized medicine to revolutionize patient outcomes and shape the future of healthcare remains an exciting frontier.

REFERENCES

Banerjee, A., Chakraborty, C., Kumar, A, Biswas, D. (2020) Emerging trends in IoT and big data analytics for biomedical and health care technologies. In *Handbook of Data Science Approaches for Biomedical Engineering*, V. E. Balas, V. K. Solanki, R. Kumar, M. Khari (Eds.) UK: Academic Press, pp. 121–152.

Bibault, J. E., Giraud, P., Housset, M., Durdux, C., Taieb, J., Berger, A., et al. (2018) Deep learning and radiomics predict complete response after neo-adjuvant chemoradiation for locally advanced rectal cancer. *Scientific Reports,* 8, 12611.

Cai, J., Lu, L., Zhang, Z., Xing, F., Yang, L, Yin, Q. (2016) Pancreas segmentation in MRI using graph-based decision fusion on convolutional neural networks. *Medical Image Computing and Computer Assisted Intervention*, 9901, 442–450.

Cheng, J., Zhang, J., Han, Y., Wang, X., Ye, X., Meng, Y. et al. (2017) Integrative analysis of histopathological images and genomic data predicts clear cell renal cell carcinoma prognosis. *Cancer Research*, 77, e91–e100. doi:10.1158/0008-5472. CAN-17-0313

Ciurtin, C., Jones, A., Brown, G., Sin, F. E., Raine, C., Manson, J. et al. (2019) Real benefits of ultrasound evaluation of hand and foot synovitis for better characterisation of the disease activity in rheumatoid arthritis. *European. Radiology*, 29(11), 6345–6354. doi:10.1007/s00330-019-06187-8.

Coelewij, L., Waddington, K. E., Robinson, G. A., Chocano, E., McDonnell, T., Farinha, F. et al. (2021) Serum metabolomic signatures can predict subclinical atherosclerosis in patients with systemic lupus erythematosus. *Atvb*, 41(4), 1446–1458, doi:10.1161/ATVBAHA.120.315321.

Nahavandi, D., Alizadehsani, R., Khosravi, A., Rajendra Acharya, U. (2022) Application of artificial intelligence in wearable devices: Opportunities and challenges. *Computer Methods and Programs in Biomedicine*, 213(1), 1–16.

Friedman, J., Hastie, T, Tibshirani, R. (2010) Regularization paths for generalized linear models via coordinate descent. *Journal of Statistical Software*, 33, 1–22. doi:10.18637/jss.v033.i01

Fortino, G., Trunfio, P. (2014) *Internet of Things Based on Smart Objects: Technology, Middleware and Applications*. United States: Springer.

Hamidinekoo, A., Denton, E., Rampun, A., Honnor, K, Zwiggelaar, R. (2018) Deep learning in mammography and breast histology, an overview and future trends. *Medical Image Analysis*, 47, 45–67.

Halicek, M., Lu, G., Little, J.V., Wang, X., Patel, M., Griffith, C.C., et al. (2017) Deep convolutional neural networks for classifying head and neck cancer using hyperspectral imaging. *Journal of Biomedical Optics*, 22(6), 60503.

Hassan, Q. F., Khan, A. R, Madani, S. A. (2017) *Internet of Things: Challenges, Advances, and Applications*. Florida: CRC Press.

Huang, Z., Zhan, X., Xiang, S., Johnson, T.S., Helm, B., Yu, C.Y., et al. (2019) SALMON: Survival analysis learning with multi-omics neural networks on breast cancer. *Frontiers in Genetics,* 10, 166.

Huang, S., Chaudhary, K, Garmire, L. X. (2017) More is better: Recent progress in multi-omics data integration methods. *Frontiers in Genetics*, 8, 84. doi:10.3389/fgene.2017.00084

Holler, J., Tsiatsis, V., Mulligan, C., Avesand, S., Karnouskos, S, Boyle, D. (2014) *From Machine-to-Machine to the Internet of Things: Introduction to a New Age of Intelligence*. United States: Academic Press.

Joyce, T., Chartsias, A, Tsaftaris, S. A. (2018) Deep multi-class segmentation without ground-truth labels. In *Proceedings of 1st Conference Medical Imaging* Deep Learnning, Paris.

Kaplan, J. (2016) *Artificial Intelligence: What Everyone Needs to Know:* What Everyone Needs To Know. New York: Oxford University Press.

Katzman, J. L., Shaham, U., Cloninger, A., Bates, J., Jiang, T. T, Kluger, Y. (2018) DeepSurv: Personalized treatment recommender system using a Cox proportional hazards deep neural network. *BMC Medical Research Methodology* 18, 24. doi:10.1186/s12874-018-0482-1

Liu, X., Faes, L., Kale, A. U., Wagner, S. K., Fu, D. J., Bruynseels, A. et al. (2019) A comparison of deep learning performance against health-care professionals in detecting diseases from medical imaging: A systematic review and meta-analysis. *Lancet Digit Health* 1(6), e271–e97. doi:10.1016/S2589–7500(19)30123-2.

Litjens, G., Kooi, T., Bejnordi, B.E., Setio, A.A.A., Ciompi, F., Ghafoorian, M.,et al. (2017) A survey on deep learning in medical image analysis. *Medical Image Analysis*, 42, 60–88.

Rajalakshmi, N. R., Saravanan, K. (2021) *Cloud-Based Intelligent Internet of Medical Things Applications for Healthcare Systems, Evolving Role of AI and IOMT in the Healthcare Market*. United States: Springer.

Rajalakshmi, N. R., Saravanan, S., Singha, A. (2023) Surplus data prediction and classification of textual-data using machine and deep learning comparative analysis. In *International Conference on Communication, Security and Artificial Intelligence (ICCSAI)* (pp. 329–334), Uttar Pradesh.

RajalakshmiN. R., Santhosh, J., Arun Pandian, J., Alkhouli, M.. (2023) Prediction of chronic heart disease using machine learning. *Smart Innovation Systems and Technologies*, 334, 177–185.

RajalakshmiN.R. (2023) Development of sign language recognition application using deep learning. *Lecture Notes in Networks and Systems*, 445, pp. 299–308.

Michalski, R. S., Carbonell, J. G, Mitchell, T. M. (2013) *Machine Learning: An Artificial Intelligence Approach*. United States: Springer.

Moeskops, P., Viergever, M.A., Mendrik, A.M., De Vries, L.S., Benders, M.J., Išgum, I. (2016) Automatic segmentation of MR brain images with a convolutional neural network. *IEEE Transactions on Medical Imaging*, 35(5), 1252–1261.

Robinson, G. A., Peng, J., Dönnes, P., Coelewij, L., Naja, M., Radziszewska, A., et al. (2020) Disease-associated and patient-specific immune cell signatures in juvenile-onset systemic lupus erythematosus: Patient Stratification using a machine-learning approach. *Lancet Rheumatology*, 2(8), e485–e96. doi:10.1016/s2665–9913(20)30168-5

Trivizakis, E., Papadakis, G. Z., Souglakos, I., Papanikolaou, N., Koumakis, L., Spandidos, D. A., et al. (2020) Artificial intelligence radiogenomics for advancing precision and effectiveness in oncologic care (Review). *International Journal of Oncology*, 57, 43–53

Trivizakis, E., Manikis, G.C., Nikiforaki, K., Drevelegas, K., Constantinides, M., Drevelegas, A., et al. (2019) Extending 2-D convolutional neural networks to 3-D for advancing deep learning cancer classification with application to MRI liver tumor differentiation. *IEEE Journal of Biomedical and Health Informatics*, 23, 923–930.

Li, Z., Wang, Y., Yu, J., Guo, Y, Cao, W. (2017) Deep learning based radiomics (DLR) and its usage in noninvasive IDH1 prediction for low grade glioma, *Science Report*, 7(1), 5467.

Witten, I. H, Frank, E. (2016) *Data Mining: Practical Machine Learning Tools and Techniques*. Netherlands: Morgan Kaufmann.

Yang, L. T., Di Martino, B, Zhang, Q. (2017) Internet of everything. *Mobile Information Systems*, Volume 2017, Article ID 8035421, London.

Zhu, Z., Albadawy, E., Saha, A., Zhang, J., Harowicz, M. R., Mazurowski, M. A. (2019) Deep learning for identifying radiogenomics associations in breast cancer. *Computers in Biology and Medicine*, 109, 85–90.

Index

Note: **Bold** page numbers refer to tables and *italic* page numbers refer to figures.

For Product Safety Concerns and Information please contact our EU
representative GPSR@taylorandfrancis.com
Taylor & Francis Verlag GmbH, Kaufingerstraße 24, 80331 München, Germany

www.ingramcontent.com/pod-product-compliance
Ingram Content Group UK Ltd.
Pitfield, Milton Keynes, MK11 3LW, UK
UKHW022315100726
473146UK00009B/482